MINUTES OF THE LEAD PENCIL CLUB

MINUTES
OF THE

PULLING THE PLUG ON THE ELECTRONIC REVOLUTION

EDITED BY BILL HENDERSON

Letters, essays, cartoons, and commentary
on how and why to live
contraption-free
in a computer-crazed world.

PUSHCART PRESS ■ WAINSCOTT, NEW YORK

The Lead Pencil Club thanks the hundreds of members who have contributed letters, essays and news items to our *Minutes*. We have attempted to clear all appropriate permissions, but if we missed you, please contact us so that we may give you proper credit in a future printing. The Club thanks the following publishers for reprint permission; (other reprint permissions were cleared with the authors.) "The Myth of Computers In The Classroom," *The New Republic*. "Virtual Students, Digital Classroom," *The Nation Co., L.P.* "A Little Cypergrouch," © 1995 *The New York Times Company*, reprinted with permission. "A Read-Only Man In An Interactive Age," reprinted with permission of *The Wall Street Journal* © 1995, Dow Jones & Co, Inc. All rights reserved. "My Son Saves Nanoseconds," © 1995, The Chronicle Publishing Co. "Cybersuicide: Quitting the Net" from *Surfing the Net* by J.C. Herz. By permission of Little, Brown and Co. "Why I Am Not Going to Buy A Computer" from *What Are People For?* by Wendell Berry © 1990 by Wendell Berry. Reprinted by permission of North Point Press, a division of Farrar, Straus and Giroux Inc. "Invalids in the Garbage of Memory" sections from *Labyrinths*, Jorge Luis Borges © 1962, 1964 New Directions Publishing Co. "Faust in the Computer Age," *The East Hampton Star*. "Scribble, Scribble, Eh Mr. Toad," © *Time*. "As It Is" from *What We Carry*, B.O.A. Editions, 1994. Cartoon permissions: "Herman," Universal Press Syndicate. Chip Bok cartoon by permission of Creators Syndicate. Drawing by D. Reilly and Agee © 1993 and 1995, *The New Yorker* Magazine Inc. "Dilbert" reprinted by permission of United Feature Syndicate Inc. All other cartoon permissions by the artists of the drawings.

This book was designed by CMYK Design, Ossining, New York.

Contents

THIS BOOK IS DEDICATED TO THE MEMORY OF

RUBE GOLDBERG (1883-1970)

Where are you when we need you most?

MINUTES OF THE LEAD PENCIL CLUB

THE LEAD PENCIL CLUB

IS THE LEAD PENCIL CLUB a put on? Are these people serious? Isn't this just some exercise in nostalgia? There's no stopping technology, right?

Our club is not only serious, it's also practical. We are not romantics, we are pragmatists. And we don't lack a sense of humor. But what computers are doing to us, our society and our children is a disaster. We are fed up with the electronic industry's hype of convenience and speed. From the Internet to television to Voice Mail to faxes to E-mail to the World Wide Web, we have had it. This nonsense has got to stop. In our *Minutes* the members of The Lead Pencil Club will tell you why and how you can yank the plug on the international electronics industry that is fattening its purse while brain-draining this civilization.

We have been informed by the industry that the march of computers and assorted gadgetry into every home, business, school, library, and pocket is "Inevitable." Microsoft's Bill Gates and his pals tell us we can't fight back because in the very near future everything will be digital. If you aren't hooked up now, your children will be illiterate and your business will be bankrupt. Worse yet, you will be cut off from the Internet and therefore from the rest of the world. In short, you will be branded a PONA (Person Of No Account, in cyberspeak).

Since the victory of gadgetry is "Inevitable," you might as well come along quietly and let us smother you in speed and convenience, says the digital mafia.

This is propaganda of the most insidious sort, threatening those of us who don't want and don't need contraptions in our lives. If you are the parent of a young child, as I am, you are assaulted with a constant advertising and media onslaught that makes you feel both guilty and terrified if you do not spring for a computer immediately. Every time your child comes home with tales of new computers and software on the desks of neighborhood kids, you imagine yourself to be both a cheapskate and crank for insisting on paper, pencil and a mind free from machine programming.

You worry that maybe all of this *is* inevitable. The stock market is in an electronic feeding frenzy. Microsoft made Bill Gates the richest man in the United States in just a few years. Steven Jobs collected a billion dollars overnight on the stock offering of his Pixar cartoon computer animation device. Businesses are desperate to stake a claim on a World Wide Web site. Millions of ordinary folk tap into the Internet for sex, info, and chat. The New York Public Library warns of a "new Dark Ages in literacy" if it can't raise the funds for computers. School boards trim teachers' salaries and slash programs so they can afford new and better computer labs. Teachers worry about the day when they will be terminated so a computer can teach and entertain the kids. No big business today will answer the phone without Voice Mail or do without E-mail. No letter can be simply mailed: It must be faxed or we are convinced it will be ignored as unimportant.

The electronic wizards tell us we can't balance our checkbooks without Quicken, research a simple fact without a CD-ROM, and we shouldn't even attempt our own handwriting—there is a software program with computer fonts that duplicate handwriting fast and neatly for a mere $39.95. Soon, so Bill Gates envisions in his instant bestseller *The Road Ahead*, we won't dare to stroll the street without a pocket PC that will be combination fax, stock market monitor, digital cash dispenser, E-mail receiver and sender, storer of our kids' photos, and—get this—a Global Positioning System that will let you know within 300 feet exactly where you are on the globe. You didn't know you needed this? Well, you do and it's "Inevitable!"

The Info Highway is paving us over.

But it gets worse. As if we needed any more proof that near absolute power corrupts nearly absolutely, the cyberspace cadets have concocted a new religion complete with a passel of techno-evangelists and their own secret lingo—plus of course, the Apocalypse right around the corner.

According to these techno-evangelists, the digital revolution is the most stunning advance in evolution since the capture of fire. They preach that nothing will be the same by the Millennium. Their gospel rejoices that we are witnessing the end of the human race. We will soon be replaced by an electronic Superbeing. This will be *The* Holocaust.

Sound a bit alarmist to you?

Here's techno-evangelist, John Perry Barlow, *Wired* magazine prophet. "When the yearning for human flesh has come to an end, what will remain? Mind may continue, uploaded into the Internet, suspended in an ecology of voltage, as ambitiously capable of self-sustenance as was that of its carbon based forebears. It is not a matter of embracing this process, it may have already embraced us, and may have in fact designated us for it in the first place."

Stewart Brand of *Whole Earth Catalogue* fame comments, "We are as gods and we might as well get good at it."

We PONAs of the Lead Pencil Club say "Enough!" We want no part of your new Apocalyptic religion, your demi-gods of Speed and Convenience. As to your ubiquitous proclamation of the forthcoming Information Age, you must be daft. We are drowning in information right now.

In our *Minutes* we encourage you to resist this gizmo juggernaut and revel in your PONA status. It's just fine to be MEAT (cyberslang for a living being). RL (Real Life) beats VL (Virtual Life) any day and the techies need a moral reeducation.

The Lead Pencil Club was founded in the middle of a December night in 1993 when this writer discovered some Luddite sentiments in Doris Grumbach's memoir *Extra Innings*. She was sick of her word processor and other gadgets and wished they would catch a virus for which there was no cure. "Why not use a lead pencil?," I asked myself. "Why not a lead pencil club, for those of us who agree with her?" The next day I called Doris with my idea and she liked it. I discovered that Henry David Thoreau was the son of a pencil maker and helped his father manufacture pencils. Indeed, it is quite probable that *Walden* was written with a pencil that Thoreau made himself. He became a Founder emeritus of our new club with his father, John Thoreau, as proprietor. Our Club was conceived in the spirit of amazed outrage tinged with humor. Our Manifesto, which starts on page 233, was mentioned in a local newspaper after I mailed it to the editor, expecting to be ignored. Instead, it appeared, and I was deluged with mail in response. I next wrote to the *New York Times* Op Ed page expecting nothing and was instead featured. The Manifesto and news of our club appeared in the Paris-based *International Herald Tribune, Time, Newsweek, The L.A. Times Book Review, Newsday,* and news services around the world. A German magazine, *Spotlight,* fashioned a one-on-one debate between Bill Gates and this writer. From all areas of the earth, letters arrived in pencil. Our correspondents wrote that it wasn't a humorous matter anymore. They were baffled, infuriated· and personally and financially injured by the onslaught of the all-pervasive electronic snake oil in their homes, offices, libraries, and schools. To them the lead pencil became a symbol of defiance at the digital colonization of this planet, just as Gandhi's spinning wheel symbolized the resistance of the people of India faced with England's imperialism. To our members it seems we are destined to be born in electronics, bathed in it all our lives and ushered out like chickens in a barn that never see the real light of day. What is now at stake is not just the oppression of omnipresent machines, it is our entire vision of what people are for. In that I was instructed by our members and to them I am thankful.

In our *Minutes*, we members of the Lead Pencil Club have engaged our brains to discover some hidden truths about the world of cyberspace. In our letters, news clips (all supplied by club members), essays and testimonials, you will discover that just behind the current blather about a free, unfettered international Internet community lurks a universal trust of business interests with billions of dollars to invest in fiber optic connections to every living room on earth. Call this trust Microsoft–QVC–Disney–Time/Warner Inc. and you have an idea of what's ahead. This new global corp will pipe in news, entertainment, commercials, and home shopping. It will be able to target market consumers. The priests of this particular cyberspace sect don't give a hoot about "mind suspended in voltage," they see consumers pinned to a map like cockroaches, sorted according to buying preferences, age, sex, and every other detail and sold to accordingly. Control is what's ahead and like TV, big bucks will control cyberspace. For a hint at the details plotted for the next decade, check your daily TV (go ahead try it). Sex, violence, and more sex. Since virtual sex (masturbation) via chats in hundreds of alt.sex groups is a major physical activity of cyberjocks, you can be sure that virtual sex will continue to be a planetary moneymaker in cyberspace.

The electronic media has managed to trivialize everything we value by now. But sex, you see, is the one instinctual life affirming joy that the commercial boys and girls have not yet managed to totally bore us with. Something about children being born from sex and the continuation of the human race, can't quite make sex trivial. So it still sells stuff.

But when this "yearning for flesh" is over and "mind is lifted up into the net" what then will we do in cyberspace? Will the new VR become a VR of the old VR and so forth into the future? I think the cyber-theologians have some figuring to do.

But I digress. This is, after all, a helpful and spirited book. Here are some of the PONAs who help you resist Global Cyber-Corp in the pages ahead: Russell Baker, Wendell Berry, Sven Birkerts, Clifford Stoll, Neil Postman, Andrei Codrescu, David Gelernter, with guest appearances by Gary Snyder, Robert Hughes, John Updike, E. Annie Proulx, Alvin Toffler, and many more.

This is, as I said, a practical volume. Our contributors will tell you what's wrong and how to fix it. We suggest that when you destroy our ideas of who we are you cancel our future. Each time you give a machine a job to do you can do yourself, you give away a part of yourself to the machine. That's not practical. If you drive instead of walk, if you use a calculator instead of your mind, you have disabled a portion of yourself. On the other hand, every time you remove a technology from your life, you discover a gift.

Of course Gates and Co. proclaim from their silicon mountain that their tools will benefit us immeasurably. That we will be elevated by their machines, not enervated. And no doubt, particularly in medical technology, they are correct.

But let's think for a second about instant information access, the centerpiece of their wizardry. As Neil Postman points out, we are already stuffed like geese with 260,000 billboards, 11,520 newspapers, 11,556 periodicals, 27,500 outlets for video rentals, 500 million radios, 100 million computers, 40,000 new books a year, and 60 million pieces of junk mail annually. Plus 98 percent of our homes have TVs and many more than one. We need more information? Get a life!

Yes, but what of the Internet's World Wide Web? Doesn't the Library of Congress have a site there? Can't I download all their books on my magic toy? No you can't and, for most books, you never will. You can window shop through library listings but most of the books you want stay in the library and you will have to go get them.

The Lead Pencil Club declares that the World Wide Web is a duck primed for slaughter. Soon, with every individual, organization and company owning a site, the World Wide Web will resemble the World Wide Yellow Pages, and you know how much fun it is to flip through the phone book.

All this and more you will learn from our contributors. You will discover what it's like to instruct Vietnamese tribesmen with computers; how electronics are being dumped on third world countries in place of necessary aid; how cyberaddiction on campus is destroying students; how old fashioned paper and binding books are just plain wonderful; how word-processors (a term to gag on) mess up your writing style; how the Internet destroys self-publishing authors; why the Swiss Army slaughtered all its passenger pigeons thanks to the digital revolution; how computers befouled the Notre Dame Cathedral organ in Paris; how the Amish isolate their telephones. You will also be instructed in the joys of chucking out your radio, and read anguished confessions of a former Disney engineering exec; plus an epistle from a teacher on an island in Micronesia with the best computer lab in the South Pacific (but no toilets). You will be reminded of the happiness of playing with children from a LPC member in a village in India, and you will be taught a simple lesson in economics from a California LPC member who discovered it's cheaper to send a letter with a 32 cent stamp than to buy a $3,200 computer and send one by E-mail.

Finally, you will be the first to read a letter from deep inside the CIA by the last CIA editor using a pencil instead of a computer, and a subversive essay from a teacher who is not about to let computer programs teach two-year-olds before their feet have hit the sandbox. In conclusion, you may agree with a New York City member who writes, "I think the Internet must be the most anti-social thing except for guns."

These writers remind us of what we are losing in our Faustian deal with electronics. But why should we have to remind people? Isn't it obvious when a computer is used to duplicate your handwriting that you surrender part of your personality found in that handwriting? Isn't it plain that when your child proudly produces a work of art employing a computer that your child has lost a precious part of her imagination surrendered to the machine program? Isn't it quite apparent that typing on-line masturbatory details to a stranger with no known gender on an Internet alt.sex group is so very, very sad? When our TVs haul up celebrities to worship for no reason except they are declared worthy of worship this week (to be replaced next week by new faces) can't we see that our gods are duds? When all reality is declared to be virtual, when we are further and further removed from our cultural moorings, why does it seem strange that some people think it is fine to blow up an office building in Oklahoma City killing 168 men, women, and children for a virtual cause? Can't we see how lost in cyberspace we are already?

SO WHAT CAN YOU DO against the reign of Gates and the gospel according to *Wired*? Lots it turns out. We have hundreds of suggestions in the pages ahead.

Personally, you can adopt Sven Birkerts' motto about electronic technology, "refuse it." At home, hang up on Voice Mail and unplug the TV. Persuade your children that they would be better off without a computer. Sit around the fire in the evening (a real fire in the fireplace) and read to them or let them read to you. Play games that use the brain, not pickle it. After years of lobbying for a computer, my own daughter recently announced she'd prefer a clarinet for Christmas. For me this was a terrific personal victory. She had played the CD-ROMs of her friends and opted instead for music she could play with her mind and spirit. Now of an evening, I play my trumpet, and she the clarinet. The TV is dark.

At work eschew E-mail; fax not; pencil and paper will produce a better response; a trip to the water cooler evokes more interesting office chatter. Haste makes waste. What's your hurry? Not so fast. Speed kills.

If you still feel lonely and cut off from web sites and Internet sex and chat try communing again with your wife or husband or real life lover. Read a book. (We include a "Book Report" at the end of these Minutes.)

Subscribe to *Plain* (P.O. Box 100, Chesterhill, Ohio 43728) a low-tech magazine that will soothe your electronically jangled soul. Take out a stack of poetry from your library, and while you are there inquire about what they did with the card catalogue. Maybe it's in the basement and they could bring it back upstairs. Here's a poem by T. S. Eliot that might offer solace for your lonely unwired self.

The endless cycle of ideas and action,
Endless invention, endless experiment
Brings knowledge of motion but not of stillness;
Knowledge of speech but not of silence;
Knowledge of words and ignorance of the word.
All our knowledge brings us nearer to our ignorance—
Where is the life we have lost in living?
Where is the Wisdom we have lost in knowledge?
Where is the Knowledge we have lost in information?
The cycles of heaven in twenty centuries bring us farther from God
And nearer to the dust.

<u>12</u>   IF YOU WANT TO JOIN our club we welcome you. Acceptance is automatic. No dues. We've got a tasteful T-shirt you can buy (the sale of these shirts, now in their fourth printing, helped pay for the publication of this book). We have a member's certificate for your wall, and from time to time I knock out a newsletter on my 1942 Royal manual typewriter. We are at: The Lead Pencil Club, P.O. Box 380, Wainscott, NY 11975 and we'd love to hear from you.

You rebel you.

BH
January 1996

PENCILS

GAIL LAWRENCE

I SHARPENED A LOT OF PENCILS while I was working on a recent book. A freshly sharpened pencil gave me the right attitude toward revising, and the act of sharpening offered me convenient breaks from my labor. I found myself not only sharpening my pencils but examining their lead, rubbing dirty smudges of it between my fingertips, and even looking at the different marks different pencils made through my hand lens. At one point I dropped the chapter I was revising and addressed my full attention to lead.

The first thing I learned is that pencil "lead" is not really lead. It is called "lead" only because until 1779 everyone thought it was merely a darker form of the familiar metal. In that year a Swedish chemist named Karl Wilhelm Scheele proved that the substance called pencil lead was actually a form of carbon. Eventually it was given a new name—graphite, from the Greek word *graphein,* meaning "to write."

Although graphite is technically a mineral, defined as an inorganic substance, it—like oil—was once living. It is derived from plant matter that has been metamorphosed by heat and pressure into almost pure carbon. It's related to both coal and diamonds, coal having been subjected to less heat and pressure, diamonds to more.

In graphite, the carbon atoms are arranged in layers, the atoms right beside each other closely linked but those above and below only loosely linked. This internal structure makes graphite tend to slip apart in thin, easily broken sheets. When you rub it, it will cover your fingers with dark little pieces of itself, which at the tip of a nicely sharpened pencil would form a dark line on a piece of paper.

Human beings had devised lots of implements to draw and write with before they discovered graphite. They used chalk, bits of burned wood, brushes made from plants or animal fur, reeds, and quills. The Egyptians, Greeks, and Romans used pieces of the metal lead to draw lines, and during the 1300s artists used thin rods of lead for artwork. But it wasn't until 1564, when a deposit of almost pure graphite, called *plumbago*, or "that which acts like lead," was unearthed in England, that pencils as we know them became a possibility.

At first people just used chunks of this pure graphite, but that got their fingers dirty. So someone decided to wrap twine around long, narrow pieces of graphite to create a primitive pencil. Next came the idea of encasing the graphite in wood, and the pencil began to take on its modern form. But these early pencils were still clumsy, messy tools that produced thick, dark, easily smeared lines. Finally, in 1795, a French inventor named Nicholas Jacques Conte, who was commissioned by Napoleon to come up with a French version of the pencil, thought of mixing moist clay with powdered graphite and baking this mixture into pencil leads. He discovered that he could create leads that would write lighter, thinner lines by adding more clay and darker, thicker lines by adding less clay.

If you'd like to explore some of the effects that are now possible with the varying proportions of clay offered by modern pencil manufacturers, buy yourself several pencils with different numbers on them. The standard number 2 is a mixture of about two-thirds graphite and one-third clay. If you like darker, broader lines, you can use a number 1, which includes less clay. For lighter, finer lines, try a number 3 or 4, which include more clay. You may also notice numbers 2½, 2⅝, and 2.5, all of which offer the same grade of lead. The numbers are different because when the Eagle Pencil Company came out with a number 2½, the numerals became part of their trademark. Other pencil companies

15

that wanted to produce the same intermediate grade of pencil, which consumers seemed to like, were prevented by law from calling their pencils 2½, so they used 2⁵⁄₁₀ and 2.5 instead.

If you want to experiment with the extremes that are available today, buy some art and mechanical drawing pencils. Art pencils rated 4B or higher produce lines darker than the number 1 writing pencil. A 6B makes soft, bold lines that looked like black crayon through my hand lens. For the other extreme, look for mechanical drawing or drafting pencils rated 4H and higher. A 6H makes exceedingly fine lines that look like spider silk next to the strokes of the 6B.

Interestingly enough, none other than Henry David Thoreau was one of the early innovators who helped to develop the kinds of pencils that are now available in this country. His father was a pencil maker, and Thoreau went to work for him when he graduated from Harvard. Because young Thoreau was dissatisfied with the family pencil, he studied European pencil-making techniques to learn how other countries made their superior products. He discovered that the Germans were adding a very special clay from just one mine in Bavaria, found a supply of this clay, and added it to his family's formula. Then he decided that the graphite should be ground finer, and all the grinding that ensued, one scholar suggests, may have exacerbated the tuberculosis that killed both Thoreau and one of his sisters at an early age.

When I'm struggling with something as frustrating as revising what I've written, I'm easily distracted. Most of my distractions make me feel guilty, but my study of pencils—especially when I learned that Thoreau, too, had taken a serious interest in them—seemed legitimate respite. After all, if I'm going to continue to write, I need lots of pencils, and sharpening all those new pencils I bought gave me an excellent attitude toward getting back to work.

I use a Poor Man's fax (a postcard) as often as possible and it gets their attention and a smile.

Phillip Grazide
Fulton, California

Americans throw away more than 12 million computers each year, adding 600 million pounds of garbage and toxic waste to over-burdened landfills.

DENVER POST

cybersex: *masturbating while sharing a fantasy on a computer Net with someone else logged on at the same time.*
virtual reality: *fully sensual "live-in" environments that we will be able to enter by hooking ourselves up to a computer. Ideally, the computer will generate an illusion indistinguishable from the real thing.*

I'll Take the Superhighway's Sideroads

TESTIMONY BY
MICHELE M. CARDELLA

RECENT TELEVISION ADS for MCI offer a frightening glimpse into the future. The oracle is Anna Pacquin, the Academy Award nominee who played a poorly parented child in the movie "The Piano." In these commercials she is again shrouded ghoulishly in black. She stands alone in a snowy field or on a rocky coast; no friends are waiting nearby for her to get back to the game, nobody is calling her to supper. Looking straight into the camera, she relays a parable about the Information Superhighway. In words as chilling as the air that veils her breath in white, she warns of a future when "every book, every movie . . . every piece of knowledge in the universe" will be transmitted electronically into our homes. "There will be no there," she cautions, "there will only be here." Surprisingly, Pacquin offers this as good news.

While I like my home well enough, I don't look forward to being held prisoner there in order to secure information and consumer goods, attend movies, concerts and theater, enjoy what used to be books and art, and take interactive video vacations. I don't want to be one of the herd signing on to E-Mail to reminisce about the days before global agoraphobia, when 30 hours a week in front of the television seemed like a lot. When kids went to school and adults went to work. When we had stores and libraries and made a friend standing in line for playoff tickets. When we had to go out there to haul it in here.

The Information Superhighway stands at the sink with steel wool in hand ready to scrape the seasoning off of life's omelet pan, making everything cooked in it overdone and less flavorful. Even simple things will be affected—like buying a newspaper. This morning at the newsstand on New York's Chambers Street, I handed Ahmed a dollar and he gave me a newspaper, a pack of badly needed tissues and wishes for a good day. I got on the subway not particularly caring that my hands were smudged with newsprint, and relieved that my access to current events didn't require batteries or sockets or even glasses, yet. And had I finished the paper before reaching my stop, I could have tried a random act of kindness and passed it to the guy who'd been reading over my shoulder since Franklin Street.

I would rather walk into my butcher shop on Bleecker Street in the Village than switch on the Home Meat Network every Monday. I like it when Frank slips my 4-year-old a slice of bologna while we wait for him to debone the chicken breasts. I enjoy hearing about his next trip to Nashville and sharing the latest Elvis sighting I spotted in a tabloid at the grocery store checkout. Unlike the Bank Channel, Frank would float me for a week if I forget my checkbook, and with more care than the Home Mating Channel, he keeps an eye open for a nice guy to fix up with my sister.

Getting out too much isn't a big problem in America. The talent, time and money involved in deploying the Information Superhighway would be better used to improve the odds that we make it home at the end of the day unharmed. True, the more we stay home, the less likely we are to fall victim to violent crime. But since the Information Superhighway and interactive technology will make obsolete the many places where we work, shop and play, unemployment and poverty will rise. So will stock in socket covers, childproof caps and non-skid floor strips, since we'll be spending our time in the place where most accidents occur.

Perhaps this would all make sense if I knew where we were rushing to. Maybe the young messenger knows. I would like to bring her inside for a cup of hot cocoa, rub her hands between mine and ask her what the hurry is. Seasoned travelers know it's the back roads that make the journey worthwhile. It's the wrong turns and side trips that turn the routine into the exceptional. Maybe she hasn't lived through an earthquake or a hurricane or an arctic air front. She might not understand that nature has its own unpredictable sense of direct access. When the power goes out, a candle is more valuable than a lamp, a can opener beats a food processor and the touch of a neighbor means salvation.

THE ELECTRONIC HIVE:
REFUSE IT

SVEN BIRKERTS

THE DIGITAL FUTURE is upon us. From our President on down, people are smitten, more than they have been with anything in a very long time. I can't open a newspaper without reading another story about the Internet, the information highway. The dollar, not the poet, is the antenna of the race, and right now the dollar is all about mergers and acquisitions: the fierce battles being waged for control of the system that will allow us, soon enough, to cohabit in the all but infinite information space. The dollar is smart. It is betting that the trend will be a juggernaut, unstoppable; that we are collectively ready to round the corner into a new age. We are not about to turn from this millennial remaking of the world; indeed, we are all excited to see just how much power and ingenuity we command. By degrees—it is happening year by year, appliance by appliance—we are wiring ourselves into a gigantic hive.

When we look at the large-scale shift to an electronic culture, looking as if at a time-lapse motion study, we can see not only how our situation has come about but also how it is in our nature that it should have. At every step—this is clear—we trade for ease. And ease is what quickly swallows up the initial strangeness of a new medium or tool. Moreover, each accommodation paves the way for the next. The telegraph must have seemed to its first users a surpassingly strange device, but its newfangledness was overridden by its usefulness. Once we had accepted the idea of mechanical transmission over distances, the path was clear for the telephone. Again, a monumental transformation: turn select digits on a dial and hear the voice of another human being. And on it goes, the inventions coming gradually, one by one, allowing the society to adapt. We mastered the telephone, the television with its few networks running black-and-white programs. And although no law required citizens to own or use either, these technologies did in a remarkably short time achieve near total saturation.

We are, then, accustomed to the process; we take the step that will enlarge our reach, simplify our communication, and abbreviate our physical involvement in some task or chore. The difference between the epoch of early modernity and the present is—to simplify drastically—that formerly the body had time to accept the graft, the new organ, whereas now we are hurtling forward willy-nilly, assuming that if a technology is connected with communications or information processing it must be good, we must need it. I never cease to be astonished at what a mere two decades have brought us. Consider the evidence. Since the early 1970s we have seen the arrival of—we have accepted, deemed all but indispensable—personal computers, laptops, telephone-answering machines, calling cards, fax machines, cellular phones,

VCRs, modems, Nintendo games, E-mail, voice mail, camcorders, and CD players. Very quickly, with almost no pause between increments, these circuit-driven tools and entertainments have moved into our lives, and with a minimum rippling of the waters, really—which, of course, makes them seem natural, even inevitable. Which perhaps they are. Marshall McLuhan called improvements of this sort "extensions of man," and this is their secret. We embrace them because they seem a part of us, an enhancement. They don't seem to challenge our power so much as add to it.

I am startled, though, by how little we are debating the deeper philosophical ramifications. We talk up a storm when it comes to policy issues—who should have jurisdiction, what rates may be charged— and there is great fascination in some quarters with the practical minutiae of functioning, compatibility, and so on. But why do we hear so few people asking whether we might not *ourselves* be changing, and whether the changes are necessarily for the good?

IN OUR TECHNOLOGICAL obsession we may be forgetting that circuited interconnectedness and individualism are, at a primary level, inimical notions, warring terms. Being "on line" and having the subjective experience of depth, of existential coherence, are mutually exclusive situations. Electricity and inwardness are fundamentally discordant. Electricity is, implicitly, of the moment—*now*. Depth, meaning, and the narrative structuring of subjectivity—these are *not* now; they flourish only in that order of time Henri Bergson called "duration." Duration is deep time, time experienced without the awareness of time passing. Until quite recently—I would not want to put a date to it—most people on the planet lived mainly in terms of duration: time not artificially broken, but shaped around natural rhythmic cycles; time bound to the integrated functioning of the senses.

We have destroyed that duration. We have created invisible elsewheres that are as immediate as our actual surroundings. We have fractured the flow of time, layered it into competing simultaneities. We learn to do five things at once or pay the price. Immersed in an environment of invisible signals and operations, we find it as unthinkable to walk five miles to visit a friend as it was once unthinkable to speak across that distance through a wire.

My explanation for our blithe indifference to the inward consequences of our becoming "wired" is simple. I believe that we are—biologically, neuropsychologically—creatures of extraordinary adaptability. We fit ourselves to situations, be they ones of privation or beneficent surplus. And in many respects this is to the good. The species is fit because it knows how to fit.

But there are drawbacks as well. The late Walker Percy made it his work to explore the implications of our constant adaptation. Over and over, in his fiction as well as his speculative essays, he asks the same basic questions. As he writes in the opening of his essay "The Delta Factor": "Why does man feel so sad in the twentieth century? Why does man feel so bad in the

Whenever I ask a computer junkie, "are we giving away our power to the magic machine?" they rarely have an answer. Perhaps our culture has reached the lemming stage.

Joan T. Weiss
Laguna Beach,
California

very age when, more than in any other age, he has succeeded in satisfying his needs and making over the world for his own use?" One of his answers is that the price of adaptation is habit, and that habit—habit of perception as well as behavior—distances the self from the primary things that give meaning and purpose to life. We accept these gifts of technology, these labor-saving devices, these extensions of the senses, by adapting and adapting again. Each improvement provides a new level of abstraction to which we accommodate ourselves. Abstraction is, however, a movement away from the natural given—a step away from our fundamental selves rooted for millennia in an awe before the unknown, a fear and trembling in the face of the outer dark. We widen the gulf, and if at some level we fear the widening, we respond by investing more of our faith in the systems we have wrought.

We sacrifice the potential life of the solitary self by enlisting ourselves in the collective. For this is finally—even more than the saving of labor—what these systems are all about. They are not only extensions of the senses; they are extensions of the senses that put us in touch with the extended senses of others. The ultimate point of the ever-expanding electronic web is to bridge once and for all the individual solitude that has hitherto always set the terms of existence. Each appliance is a strand, another addition to the virtual place wherein we will all find ourselves together. Telephone, fax, computer networks, E-mail, interactive television—these are the components out of which the hive is being built. The end of it all, the *telos*, is a kind of amniotic environment of impulses, a condition of connectedness. And in time—I don't know how long it will take—it will feel as strange (and exhilarating) for a person to stand momentarily free of it as it feels now for a city dweller to look up at night and see a sky full of stars.

WHETHER THIS SOUNDS dire or not depends upon your assumptions about the human condition—assumptions, that is, in the largest sense. For those who ask, with Gauguin, "Who are we? Why are we here? Where are we going?"—and who feel that the answering of those questions is the grand mission of the species—the prospect of a collective life in an electronic hive is bound to seem terrifying. But there are others, maybe even a majority, who have never except fleetingly posed those same

It seems to me the book has not just aesthetic values—the charming little clothy box of the thing, the smell of the glue, even the print, which has its own beauty. But there's something about the sensation of ink on paper that is in some sense a thing, a phenomenon rather than an epiphenomenon. I can't break the association of electric trash with the computer screen. Words on the screen give the sense of being just another passing electronic wriggle.

JOHN UPDIKE

I took my computer to the rifle range and blew it through with holes!
    P.S. I used a flintlock rifle.

Dr. Stephen Clark
Wayne State
University
Ogden, Utah

27

questions, who have repressed them so that they might "get on," and who gravitate toward that life because they see it as a way of vanquishing once and for all the anxious gnawings they feel whenever any intimations of depth sneak through the inner barriers.

My core fear is that we are, as a culture, as a species, becoming shallower; that we have turned from depth—from the Judeo-Christian premise of unfathomable mystery—and are adapting ourselves to the ersatz security of a vast lateral connectedness. That we are giving up on wisdom, the struggle for which has for millennia been central to the very idea of culture, and that we are pledging instead to a faith in the web.

There is, finally, a tremendous difference between communication in the instrumental sense and communion in the affective, soul-oriented sense. Somewhere we have gotten hold of the idea that the more all-embracing we can make our communications networks, the closer we will be to that connection that we long for deep down. For change us as they will, our technologies have not yet eradicated that flame of a desire—not merely to be in touch, but to be, at least figuratively, embraced, known and valued not abstractly but in presence. We seem to believe that our instruments can get us there, but they can't. Their great power is all in the service of division and acceleration. They work in—and create—an unreal time that has nothing to do with the deep time in which we thrive: the time of history, tradition, ritual, art, and true communion.

The proselytizers have shown me their vision, and in my more susceptible moods I have felt myself almost persuaded. I have imagined what it could be like, our toil and misery replaced by a vivid, pleasant dream. Fingers tap keys, oceans of fact and sensation get downloaded, are dissolved through the nervous system. Bottomless wells of data are accessed and manipulated, everything flowing at circuit speed. Gone the rock in the field, the broken hoe, the grueling distances. "History," said Stephen Daedalus, "is a nightmare from which I am trying to awaken." This may be the awakening, but it feels curiously like the fantasies that circulate through our sleep. From deep in the heart I hear the voice that says, "Refuse it."

Amid all his earnest patter about personal computers being tools of empowerment, Bill Gates, the billionaire chairman of Microsoft, offered CNN's Larry King three words last week that finally touched on what all the fuss over a software program was really about.

"Software is cool," Mr. Gates said.

NEW YORK TIMES

I've been wedded to the yellow #2 since I was a kid. . . . It has served me well through school and for the past 29 years as an artist. Of all the art supplies there is none that can hold a candle to a soft #2 and a plain piece of white bond paper.

Don Lubov
Southold, New York

29

ADDING MEMORY

ANDREI CODRESCU

THE OTHER DAY, a friend of mine was explaining how she had to move these pixels around her computer and had to add 20 megabytes of memory to handle the operation. I had the disquieting thought that all this memory she was adding had to come from somewhere. Maybe it was coming from me, because I couldn't remember a thing that day. And then it became blindingly obvious: *all* the memory that everybody keeps adding to their computers comes from people. Nobody can remember a damn thing. Every time somebody adds some memory to their machine thousands of people forget everything they knew. Americans are singularly devoid of memory these days. We don't remember where we came from, who raised us, when our wars used to be, what happened last year, last month, or even last week. Schoolchildren remember practically nothing. I take the Greyhound bus every week and I swear half of the people on there don't know where they got on or where they are supposed to get off. The explanation is simple: computer companies are stealing human memory to stuff their hard drives. Greyhound, I believe, has some kind of contract with IBM, to steal the memory of everyone riding the bus. They are probably connected by a cable or something: every hundred miles, poof, her five hundred megabytes get sucked out of the passengers' brains. The computers' thirst for memory is bottomless:

the more they suck the more they need. Eventually, we will all be walking around with a glazed look in our eyes, trying to figure out who it is we live with. Then we'll forget our names and addresses and just be milling around trying to remember them. The only thing visible about us will be these cables sticking out of our behinds, feeding the scraps of our memory to Computer Central somewhere in Oblivion, USA. I think it's time for all these money-sucking companies to start some kind of system to feed and shelter us when we forget how to eat, walk, and sleep.

———— ✐ ————

Americans are suckers for utopian promises. They have been ever since the Puritans invented the idea of radical newness, in the 17th century. We will look back on what is now claimed for the information superhighway and wonder how we ever psyched ourselves into believing all that bulldust about social fulfillment through interface and connectivity. But by then we will have some other fantasy to chase, its approaches equally lined with entrepreneurs and flacks, who will be its main beneficiaries.

ROBERT HUGHES

I work part-time as a cashier and when the customer owes me $1.98 and gives me $2.00 I find myself looking at the screen to see how much change I should give her. Pitiful, isn't it?

Suzanne Mack
Norristown, Pennsylvania

POWERLESS POWER:
HOW BARBARIANS DEFEATED
TECHNOLOGY

SANJA BRIZIĆ ILIĆ

A FEW YEARS AGO in the then-Socialist Federal Republic of Yugoslavia, the country with the worst roads in Europe, many thought it more important to get on what would become known as the information superhighway than to build driveable highways of the old-fashioned asphalt kind. WordPerfect workshops mushroomed, entrepreneurs peddled poorly assembled IBM clones, and trendy columnists insisted that catapulting Yugoslavs into cyberspace (they called it "international communications network") was essential to the federation's future. The 1990s gospel preached that modern technology will make the planet a better place to live, bringing political adversaries and economic competitors together, united in a global electronic brotherhood.

But instead of speeding into the bright, laptop-equipped future, the whole region sank into the murky, terror-filled past. Yugoslavia broke apart, Serbia invaded its neighbors' territories, and a war erupted. From rushing towards the 21st century, millions of people were suddenly sucked back into the Dark Ages. Business executives who couldn't live without a fax machine found themselves huddled in damp basements lit only by candles. Cramped spaces were shared by neighbors who'd never exchanged a word before. No phones (cellular or other) rang, no neon signs lit the cities; in the darkness, only the roar of enemy bombers ripped the dead silence. And time ticked by slowly. People had forgotten how to spend time without pre-packaged electronic entertainment, or how to communicate unaided by devices and gadgets. Those who wanted to voice out their feelings or record their thoughts didn't know how to do it without the machines that ran on electric power. The species evolved to depend on the technology now had to regress to survive in a pre-technological environment.

And as the war dragged on, the forgotten skills came back. Without electricity and gas, elaborate meals were prepared on ancient wood-burning stoves. Kids discovered that the stories told by their old geezer neighbors were actually much more exciting than Nintendo. Family members became interested in each other's lives. Journalists and writers took pencils into their keyboard-addicted hands and churned out pages and pages of articles, essays, short stories, even novels. From tunnels and tents, fields and frontlines, reports and ideas kept pouring to the outside world, scribbled on whatever paper was available and delivered by live human couriers rather than E-mail. With all the devastation and the horrors, the war—the ultimate reality check—brought one good thing to Croatians and Bosnians: The barbarian invasion showed them what a powerless power technology is, and they learned how to survive without it. They *did* survive without it.

Would we?

Will we?

———— ✐ ————

Perhaps as a result of our rampant cellularization, as we fax and phone and XMIT paperless messages through air and ether, we paradoxically crave, it seems, written proof that we still exist as literate creatures. Notepaper condolences, no matter how awkwardly expressed, offer more surcease than printed cards. Vast sums are paid by chirophiles for manuscripts of eminent writers who still work in longhand, or who at least hunt and peck and cut and paste. Museums are not about to display the floppies of any contemporary Flaubert.

EDMUND MORRIS
THE NEW YORKER

After Star Wars *and its sequels made hundreds of millions of dollars, the filmmaker George Lucas built himself a self-contained, state-of-the-art studio with all the most advanced special-effects technology. The core of this studio, both architecturally and spiritually, according to Lucas himself is the books in his library.*

JOE DAVID BELLAMY

Why I Am Not Going to Buy a Computer

TESTIMONY BY
WENDELL BERRY

LIKE ALMOST EVERYBODY else, I am hooked to the energy corporations which I do not admire. I hope to become less hooked to them. In my work, I try to be as little hooked to them as possible. As a farmer, I do almost all of my work with horses. As a writer, I work with a pencil or a pen and a piece of paper.

My wife types my work on a Royal standard typewriter bought new in 1956 and as good now as it was then. As she types, she sees things that are wrong and marks them with small checks in the margins. She is my best critic because she is the one most familiar with my habitual errors and weaknesses. She also understands, sometimes better than I do, what *ought* to be said. We have, I think, a literary cottage industry that works well and pleasantly. I do not see anything wrong with it.

A number of people, by now, have told me that I could greatly improve things by buying a computer. My answer is that I am not going to do it. I have several reasons, and they are good ones.

The first is the one I mentioned at the beginning. I would hate to think that my work as a writer could not be done without a direct dependence on strip-mined coal. How could I write conscientiously against the rape of nature if I were, in the act of writing, implicated in the rape? For the same reason, it matters to me that my writing is done in the daytime, without electric light.

I do not admire the computer manufacturers a great deal more than I admire the energy industries. I have seen their advertisements, attempting to seduce struggling or failing farmers into the belief that they can solve their problems by buying yet another piece of expensive equipment. I am familiar with their propaganda campaigns that have put computers into public schools in need of books. That computers are expected to become as common as TV sets in "the future" does not impress me or matter to me. I do not own a TV set. I do not see that computers are bringing us one step nearer to anything that does matter to me: peace, economic justice, ecological health, political honesty, family and community stability, good work.

What would a computer cost me ? More money, for one thing, than I can afford, and more than I wish to pay to people whom I do not admire. But the cost would not be just monetary. It is well understood that technological innovation always requires the discarding of the "old model"—the "old model" in this case being not just our old Royal standard, but my wife, my critic, my closest reader, my fellow worker. Thus (and I think this is typical of present-day technological innovation), what would be superseded would be not only something, but somebody. In order to be technologically up-to-date as a writer, I would have to sacrifice an association that I am dependent upon and that I treasure.

My final and perhaps my best reason for not owning a computer is that I do not wish to fool myself. I disbelieve, and therefore strongly resent, the assertion that I or anybody else could write better or more easily with a computer than with a pencil. I do not see why I should not be as scientific about this as the next fellow: when somebody has used a computer to write work that is demonstrably better than Dante's, and when this better is demonstrably attributable to the use of a

You don't have to become a wreck before knowing that nothing—absolutely nothing can replace life—not even the best of technology. In India, fortunately, 90% of the population—as they live mostly in rural areas and can't afford laptop computers—have always depended on the good old postal system (ours is the best in the world I am proud to say, given the size of the population) to keep in touch with people. (They cannot afford phones also, and what the hell can one say on the phone?) I am very happy to say that we are not slaves to the box and our TV is on just an hour or so every day, mainly for news.

How can one live a life through second hand experiences? Is there better joy in life than to look at rain, play with a child?

computer, then I will speak of computers with a more respectful tone of voice, though I still will not buy one.

To make myself as plain as I can, I should give my standards for technological innovation in my own work. They are as follows:

1. The new tool should be cheaper than the one it replaces.
2. It should be at least as small in scale as the one it replaces.
3. It should do work that is clearly and demonstrably better than the one it replaces.
4. It should use less energy than the one it replaces.
5. If possible, it should use some form of solar energy, such as that of the body.
6. It should be repairable by a person of ordinary intelligence, provided that he or she has the necessary tools.
7. It should be purchasable and repairable as near to home as possible.
8. It should come from a small, privately owned shop or store that will take it back for maintenance and repair.
9. It should not replace or disrupt anything good that already exists, and this includes family and community relationships.

1987

AFTER THE FOREGOING essay, first published in the *New England Review and Bread Loaf Quarterly*, was reprinted in *Harper's*, the *Harper's* editors published the following letters in response and permitted me a reply.

W.B.

Letters

Wendell Berry provides writers enslaved by the computer with a handy alternative: Wife—a low-tech energy-saving device. Drop a pile of handwritten notes on Wife and you get back a finished manuscript, edited while it was typed. What computer can do that? Wife meets all of Berry's uncompromising standards for technological innovation: she's cheap, repairable near home, and good for the family structure. Best of all, Wife is politically correct because she breaks a writer's "direct dependence on strip-mined coal."

History teaches us that Wife can also be used to beat rugs and wash clothes by hand, thus eliminating the need for the

We have forgotten the simple joys of life. Come to India and see how even the poorest of the poor are happy playing with their babies, giving them so much love and affection; and how they enjoy the company of friends. They laugh much more than the rich. Money can definitely buy comfort but not joy—I am happy you and your friends have at last found out about the limitations of computers.

If you want to know more come to India—the real India is villages, small towns—even big cities. We are poor, yes, but mostly we know how to live a simple, joyous life close to Mother Earth.

Kasturi Swami
Kyderabad, India

Every time we take away a technology, we find a gift underneath.

PLAIN

39

vacuum cleaner and washing machine, two more nasty machines that threaten the act of writing.

Gordon Inkeles
Miranda, Calif.

I HAVE NO QUARREL with Berry because he prefers to write with pencil and paper; that is his choice. But he implies that I and others are somehow impure because we choose to write on a computer. I do not admire the energy corporations, either. Their shortcoming is not that they produce electricity but how they go about it. They are poorly managed because they are blind to long-term consequences. To solve this problem, wouldn't it make more sense to correct the precise error they are making rather than simply ignore their product? I would be happy to join Berry in a protest against strip mining, but I intend to keep plugging this computer into the wall with a clear conscience.

James Rhoads
Battle Creek, Mich.

I ENJOYED READING Berry's declaration of intent never to buy a personal computer in the same way that I enjoy reading about the belief systems of unfamiliar tribal cultures. I tried to imagine a tool that would meet Berry's criteria for superiority to his old manual typewriter. The clear winner is the quill pen. It is cheaper, smaller, more energy-efficient, human-powered, easily repaired, and non-disruptive of existing relationships.

Berry also requires that this tool must be "clearly and demonstrably better" than the one it replaces. But surely we all recognize by now that "better" is in the mind of the beholder. To the quill pen aficionado, the benefits obtained from elegant calligraphy might well outweigh all others.

I have no particular desire to see Berry use a word processor; if he doesn't like computers, that's fine with me. However, I do object to his portrayal of this reluctance as a moral virtue. Many of us have found that computers can be an invaluable tool in the fight to protect our environment. In addition to helping me write, my personal computer gives me access to up-to-the-minute reports on the workings of the EPA

The country may be moving in the direction of a purer democracy than anything the ancient Greeks envisioned.

It promises to be a fiasco.

Opinion polls and focus groups are Stone Age implements in the brave new world of interactivity just down the communications superhighway. Imagine an ongoing electronic plebiscite in which millions of Americans will be able to express their views on any public issue at the press of a button. Surely nothing could be a purer expression of democracy. Yet nothing would have a more paralyzing impact on representational government.

Was not the idea, after all, that we would elect citizen representatives to go about the business of governance for a set period while we voters went about our own business?

TED KOPPEL

41

and the nuclear industry. I participate in electronic bulletin boards on which environmental activists discuss strategy and warn each other about urgent legislative issues. Perhaps Berry feels that the Sierra Club should eschew modern printing technology, which is highly wasteful of energy, in favor of having its members hand-copy the club's magazines and other mailings each month?

Nathaniel S. Borenstein
Pittsburgh, Pa.

THE VALUE of a computer to a writer is that it is a tool not for generating ideas but for typing and editing words. It is cheaper than a secretary (or a wife!) and arguably more fuel-efficient. And it enables spouses who are not inclined to provide free labor more time to concentrate on *their* own work.

We should support alternatives both to coal-generated electricity and to IBM-style technocracy. But I am reluctant to entertain alternatives that presuppose the traditional subservience of one class to another. Let the PCs come and the wives and servants go seek more meaningful work.

Toby Koosman
Knoxville, Tenn.

BERRY ASKS how he could write conscientiously against the rape of nature if in the act of writing on a computer he was implicated in the rape. I find it ironic that a writer who sees the underlying connectedness of things would allow his diatribe against computers to be published in a magazine that carries ads for the National Rural Electric Cooperative Association, Marlboro, Phillips Petroleum, McDonnell Douglas, and yes, even Smith-Corona. If Berry rests comfortably at night, he must be using sleeping pills.

Bradley C. Johnson
Grand Forks, N.D.

Wendell Berry Replies:

The foregoing letters surprised me with the intensity of the feelings they expressed. According to the writers' testimony, there is nothing wrong with their computers; they are utterly satisfied with them and all that they stand for. My correspondents are certain that 1 am wrong and that I am, moreover, on the losing side, a side already

I don't use pencils myself to write with—they are more fun to eat. However, I am fully in sympathy with your aims and objectives.

Farley Mowat
Ontario, Canada

I got flamed for the first time a couple of months ago. To flame, according to Que's Computer User's Dictionary, *is "to lose one's self-control and write a message that uses derogatory, obscene, or inappropriate language." Flaming is a form of speech that is unique to on-line communication, and it is one of the aspects of life on the Internet that its promoters don't advertise, just as railroad companies didn't advertise the hardships of the Great Plains to the pioneers whom they were hoping to transport out there. My flame arrived on a windy* ✏

43

relegated to the dustbin of history. And yet they grow huffy and condescending over my tiny dissent. What are they so anxious about?

I can only conclude that I have scratched the skin of a technological fundamentalism that, like other fundamentalisms, wishes to monopolize a whole society and, therefore, cannot tolerate the smallest difference of opinion. At the slightest hint of a threat to their complacency, they repeat, like a chorus of toads, the notes sounded by their leaders in industry. The past was gloomy, drudgery-ridden, servile, meaningless, and slow. The present, thanks only to purchasable products, is meaningful, bright, lively, centralized, and fast. The future, thanks only to more purchasable products, is going to be even better. Thus consumers become salesmen, and the world is made safer for corporations.

I am also surprised by the meanness with which two of these writers refer to my wife. In order to imply that I am a tyrant, they suggest by both direct statement and innuendo that she is subservient, characterless, and stupid—a mere "device" easily forced to provide meaningless "free labor." I understand that it is impossible to make an adequate public defense of one's private life, and so I will only point out that there are a number of kinder possibilities that my critics have disdained to imagine: that my wife may do this work because she wants to and likes to; that she may find some use and some meaning in it; that she may not work for nothing. These gentlemen obviously think themselves feminists of the most correct and principled sort, and yet they do not hesitate to stereotype and insult, on the basis of one fact, a woman they do not know. They are audacious and irresponsible gossips.

In his letter, Bradley C. Johnson rushes past the possibility of sense in what I said in my essay by implying that I am or ought to be a fanatic. That I am a person of this century and am implicated in many practices that I regret is fully acknowledged at the beginning of my essay. I did not say that I proposed to end forthwith all my involvement in harmful technology, for I do not know how to do that. I said merely that I want to limit such involvement, and to a certain extent I do know how to do that. If some technology does damage to the world—as two of the above letters seem to agree that it does—then why is it not reasonable, and indeed moral, to try to limit one's use of that technology? *Of course,* I think that I am right to do this.

Friday morning. I got to work at nine, removed my coat, plugged in my PowerBook, and, as usual, could not resist immediately checking my E-mail. I saw I had a message from a technology writer who does a column about personal computers for a major newspaper, and whom I knew by name only. I had recently published a piece about Bill Gates, the chairman of Microsoft, about whom this person has also written, and as I opened his E-mail to me it was with the pleasant expectation of getting feedback from a colleague. Instead, I got:

Crave THIS, asshole.
Listen, you toadying dipshit scumbag . . .

JOHN SEABROOK
THE NEW YORKER

45

I would not think so, obviously, if I agreed with Nathaniel S. Borenstein that "'better' is in the mind of the beholder." But if he truly believes this, I do not see why he bothers with his personal computer's "up-to-the-minute reports on the workings of the EPA and the nuclear industry" or why he wishes to be warned about "urgent legislative issues." According to his system, the better in a bureaucratic, industrial, or legislative mind is as good as the "better" in his. His mind apparently is being subverted by an objective standard of some sort, and he had better look out.

Borenstein does not say what he does after his computer has drummed him awake. I assume from his letter that he must send donations to conservation organizations and letters to officials. Like James Rhoads, at any rate, he has a clear conscience. But this is what is wrong with the conservation movement. It has a clear conscience. The guilty are always other people, and the wrong is always somewhere else. That is why Borenstein finds his "electronic bulletin board" so handy. To the conservation movement, it is only production that causes environmental degradation; the consumption that supports the production is rarely acknowledged to be at fault. The ideal of the run-of-the-mill conservationist is to impose restraints upon production without limiting consumption or burdening the consciences of consumers.

But virtually all of our consumption now is extravagant, and virtually all of it consumes the world. It is not beside the point that most electrical power comes from strip-mined coal. The history of the exploitation of the Appalachian coal fields is long, and it is available to readers. I do not see how anyone can read it and plug in any appliance with a clear conscience. If Rhoads can do so, that does not mean that his conscience is clear; it means that his conscience is not working.

To the extent that we consume, in our present circumstances, we are guilty. To the extent that we guilty consumers are conservationists, we are absurd. But what can we do? Must we go on writing letters to politicians and donating to conservation organizations until the majority of our fellow citizens agree with us? Or can we do something directly to solve our share of the problem?

I am a conservationist. I believe wholeheartedly in putting pressure on the politicians and in maintaining the

I work at a large university library. When I started my job twenty years ago, it was a nice profession. You typed the new books on cataloging cards and did all research with the help of books. Four years ago the new era began. We have to type new books into a computer, which breaks down often. Research is done more and more with CD-ROMs and other computer systems. . . . I feel more like a computer operator now than a librarian. My eyesight is getting worse and I suffer from headaches.

Felicitas Machimar
Mainz, Germany

It won't stop until everyone knows everything at once. What happens next? Catatonia.

MARVIN BELL

conservation organizations. But I wrote my little essay partly in distrust of centralization. I don't think that the government and the conservation organizations alone will ever make us a conserving society. Why do I need a centralized computer system to alert me to environmental crises? That I live every hour of every day in an environmental crisis I know from all my senses. Why then is not my first duty to reduce, so far as I can, my own consumption?

Finally, it seems to me that none of my correspondents recognizes the innovativeness of my essay. If the use of a computer is a new idea, then a newer idea is not to use one.

——— ✐ ———

For consumers who are devoted to family values but can't seem to find the time to start a family, Quality Video of Minneapolis has produced "Video Baby."

This 30-minute tape shows two stunt infants doing what babies tend to do, like crawl around the house, play with a rattle, take a bubble bath, and turn lunch into a complete mess. There's no narrator (and no cleanup), so once the tape is in the VCR there's nothing to come between the viewer and the baby but the off button.

NEW YORK TIMES MAGAZINE

Writing, particularly poetry, is for me a visceral, sensuous experience. I need to feel the works that I am writing. I cannot allow a machine to intrude.

Carole Henkoff
Upper Montclair, New Jersey

A LITTLE CYBER GROUCH

RUSSELL BAKER

DOES YOUR BLOOD run cold, friend, when you read about the glories of "cyberspace"? Do you have to repress a shriek of protest every time you hear or read or think about "the information highway"?

If so, it means you are an old stick-in-the-mud and are doomed to end up in the dustbin of history unless you surrender immediately and come along quietly into the age of electronics amok.

As a devout reactionary, I naturally despise what these zealous engineers propose to do to us, but cruel experience reminds me it is foolish to oppose them when they are in the heat of re-inventing the world.

My distaste for this latest creative onset begins with petty, unworthy, whining objections. Why, for instance, must they refer to what is being advertised as a magical, irresistible electronic playground as "cyberspace"?

People capable of afflicting anybody, anything or anyplace with a name like "cyberspace" surely cannot have the spiritual and esthetic delicacy essential to creation of a magical, irresistible playground, can they?

All right, call it a captious quibble, but if you were given a choice of places to spend a month, which name would you select—Tuscany or Cyberspace?

As for "the information highway," sometimes called "the information superhighway," the underlying assumption strikes me as fatally defective. The modern world is not dying for want of more information. Quite the opposite; its plight is too much information. It is being battered senseless, then buried under avalanches of information.

Day and night it is assaulted by a ceaseless flow of information. Often so much information arrives so swiftly that no one can digest it, make sense of it or judge whether it's information worth having.

The national love of gadgetry is involved here. The prospect of hundreds and hundreds of TV channels emptying into our minds, of movies pumping into our eyeballs through the telephone while incoming messages are depleting our fax-paper supply and our computer is talking to the bank and paying the gas bill is a horror reminiscent of the Mickey Mouse sequence in "Fantasia," in which the magically activated water buckets cannot be restrained in their determination to drown the world. Our love of gadgets, however, makes us see it as a delight.

Already people who once walked abroad on the great green earth and breathed the outdoor air now sit glued through the night to their electronic machines, chatting it up with similarly afflicted cyberspaceniks around the world.

All this is being promoted, most notably by Vice President Gore, as a blessing for humanity. And who is to argue with a Vice President?

Still, considering only that part of humanity that is American, it is hard to see how it is going to bless the substantial part of the population that (a) can't afford the machinery and (b) lacks the know-how to make it work.

Many high schools regularly graduate their young so innocent of computer knowledge that they have never worked a keyboard. This considerable part of the population is already going to have trouble avoiding the dustbin of history. The advent of Cyberspace Triumphant is likely to cripple it more thoroughly.

Holding itself together as a nation is already becoming difficult for the United States. The trend everywhere is toward slicing the country into slivers.

Congress, suddenly uneasy with the Union, tries to give power back to single states. Ethnic groups once content to parade one day a year now insist on year-round awareness of their tribal identities. A country that once insisted everybody be "American without a hyphen" is now restoring its hyphens.

The famous Kerner Report's prediction that increasingly divisive racial and economic-group differences would turn the United States into "two nations" looks more accurate with each passing year.

Mr. Gore apparently sees a happier future in which the good old one nation indivisible will go through life with a laptop on every knee. Let us hope so. The mood of the prevailing half of the country as expressed in the election just past and by the present Congress does not, however, seem to bode well for cyberspace for all.

The techno-freaks are trying to sell us a further corruption of communication which would benefit mainly themselves.

The other evening I said to an E-mail enthusiast, "I want to look the other guy in the eye." That may be too much to ask nowadays, but the Lead Pencil Club certainly seems to be on the right retrograde track. It may even rescue civilization as we used to know it.

Robert L. Hermann
Rosemont,
Pennsylvania

Did you know that 12,310 articles about technology and business are generated each day? It's the familiar "information explosion" dilemma. As the amount of data escalates, it seems impossible to absorb it all.

Fortunately, a solution is at hand . . .

**FROM A SALES PITCH
FOR AN ON-LINE
NEWSLETTER**

You Can't Live on the Internet

CLIFFORD STOLL

NEWSPAPER UNDER MY arm, I'm pushing a baby buggy—one with rubber bumpers—to meet a friend for coffee. Later, I'll take my infant daughter, Zoe Rose, to the library. What's strange about that?

What's strange is that I'm walking. You see, I *could* reach my friend over the Internet. We'd exchange E-mail or chat over the modem. But I prefer to meet face to face. Also strange: I could get my news from a computer network, yet I'm toting a newspaper.

My friends know me as a computer jock, one of the original nerds. And, until recently, I was. Seven years ago, I caught a hacker on the Internet. He had broken into military systems, stolen information and sold it to the Soviet KGB. After a year-long chase, I nailed him; later, he was convicted of espionage.

Today, with six computers and plenty of network links, I consider myself well-connected. Still, I'm amazed by all the attention paid to the Internet.

You've probably heard how the Internet will replace newspapers, magazines and books. We'll find online libraries and multimedia classrooms. Business and government will flock to the digital world. It'll become a part of our personal lives.

Well, I don't believe it.

Gradually, I've come to realize that the computer diverts our attention from more important things: friends, family, neighborhood. Yes, I'm constantly tempted to surf the 'Net, but I know that an afternoon hike in the forest brings more satisfaction than my modem ever could deliver.

For one thing, the Internet isn't intimate. When I'm online, there's no way to raise my eyebrow quizzically or show off my daughter's dimples. Electronic mail, so easy to compose and send, lacks the warmth of handwritten letters.

And if the 'Net is so great at delivering instant news and magazines, why am I toting this newspaper? Because newsprint is friendly. Headlines direct my attention to the most important stories. I know at a glance whether a column is worth reading. The paper's peppered with pictures and charts, completely missing from files downloaded from the 'Net. Then, too, newspapers are cheap—if I forget my copy at the café, nobody will steal it. Try that with your laptop computer.

Now that I spend less time with a modem, I'm discovering wonders in my own neighborhood. With a fresh eye, I watch kids playing chess on funny green stools. There's a real newsstand with magazines. A sky with constellations.

Sure; I could play chess over the 'Net and download stories and astronomy photos. But they're pallid imitations of the real thing. Reality is far sweeter.

Once, television promised to bring the finest entertainment into our homes. Instead we find a cultural wasteland. In the same way, I suspect the ballyhooed information highway will deliver a glitzy, non-existent world in which important aspects of human interaction are relentlessly devalued.

As I push baby Zoe, I wonder about the schools she'll attend. Will my daughter be plopped before some multimedia system, complete with "edutainment" software? Will her classroom become an interactive video game? Or will a real, caring teacher show her the joys and risks of our world? Will her library be little more than a row of personal computers connected to CD-ROM readers? Or will Zoe discover real books and magazines, as well as a lively librarian who reads stories on Saturday mornings?

I'M SURE ZOE will learn to use word processors and spreadsheets. After all, these are nearly universal in today's businesses. But I wonder if she'll appreciate more mundane—but far more essential—skills like carburetor repair, plumbing, and rolling out biscuits. And what multimedia program could teach her to do unto others as she would have others do unto her?

Two decades from now, when my daughter is 20½, I picture myself reading handwritten letters from college, not cold E-mail greetings. Better yet, we'll talk about old times over a cup of coffee, reveling in the unfaded love and respect between father and daughter. Given access to the entire Internet, who could download a more gratifying message?

WHAT THE INTERNET hucksters won't tell you is that the Internet is an ocean of unedited data, without any pretense of completeness. Lacking editors, reviewers or critics, the Internet has become a wasteland of unfiltered data. You don't know what to ignore and what's worth reading. Logged onto the World Wide Web, I hunt for the date of the Battle of Trafalgar. Hundreds of files show up, and it takes 15 minutes to unravel them—one's a biography written by an eighth grader, the second is a computer game that doesn't work and the third is an image of a London monument. None answers my question, and my search is periodically interrupted by messages like, "Too many connections, try again later."

My appropriate skills include good familiarity with lead pencils, knowledge of techniques for sharpening LPs, cleaning erasers, multicolor LP sets, tips on choosing the ideal paper medium for LPs to achieve optimum results, using LPs to produce cemetery "rubbings," how to choose the ideal lead hardness to suit the job, knowing when to use a mechanical pencil and when to avoid it, how to buy pencils at government auctions at tremendous savings, famous pencils from our past, etc.

And that's not all. For the unconvinced and skeptics there are many horror stories about leaking ballpoint pens, fountain pen disasters down through the years, and so on, not to mention the numerous recorded disasters related to hard disk crashes, power failures before data were saved, etc. ✏

57

Won't the Internet be useful in governing? Internet addicts clamor for government reports. But when Andy Spano ran for county executive in Westchester County, N.Y., he put every press release and position paper onto a bulletin board. In that affluent county, with plenty of computer companies, how many voters logged in? Fewer than 30. Not a good omen.

Then there are those pushing computers into schools. We're told that multimedia will make schoolwork easy and fun. Students will happily learn from animated characters while taught by expertly tailored software. Who needs teachers when you've got computer-aided education? Bah. These expensive toys are difficult to use in classrooms and require extensive teacher training. Sure, kids love videogames—but think of your own experience: can you recall even one educational filmstrip of decades past? I'll bet you remember the two or three great teachers who made a difference in your life.

Then there's cyberbusiness. We're promised instant catalog shopping— just point and click for great deals. We'll order airline tickets over the network make restaurant reservations and negotiate sales contracts. Stores will become obsolete. So how come my local mall does more business in an afternoon than the entire Internet handles in a month? Even if there were a trustworthy way to send money over the Internet—which there isn't—the network is missing a most essential ingredient of capitalism: salespeople.

What's missing from this electronic wonderland? Human contact.

Discount the fawning techno-burble about virtual communities. Computers and networks isolate us from one another. A network chat line is a limp substitute for meeting friends over coffee. No interactive multimedia display comes close to the excitement of a live concert. And who'd prefer cybersex to the real thing? While the Internet beckons brightly, seductively flashing an icon of knowledge-as-power, this nonplace lures us to surrender our time on earth. A poor substitute it is, this virtual reality where frustration is legion and where—in the holy names of Education and Progress— important aspects of human interactions are relentlessly devalued.

"Makes you wonder how we ever managed without it."

And then there is the well documented therapeutic value of working with a lead pencil. Many users compare it to the comforts associated with favorite sweaters, broken-in shoes, and chicken dumplings. Lives have been saved, careers salvaged, marriages renewed via the lead pencil . . . Well maybe I did get caught up in this . . .

MICHAEL JAQUISH
STONE MOUNTAIN,
GEORGIA

59

Invalids in the Garbage of Memory

PETER GRENQUIST

IN 1942 the Argentine author Jorge Luis Borges imagined a 19th century crippled prodigy, Ireneo Funes, endowed with infallible perception and memory. Funes dwelt in a darkened room to limit the number of impressions assailing his limitless comprehension. Borges wrote:

> "We, in a glance, perceive three wine glasses on the table; Funes saw all the shoots, clusters, and grapes of the vine. He remembered the shapes of the clouds in the south at dawn on the 30th of April of 1882, and he could compare them in his recollection with the marbled grain in the design of a leather-bound book which he had seen only once, and with the lines in the spray which an oar raised in the Rio Negro on the eve of the battle of the Quebracho. These recollections were not simple; each visual image was linked to muscular sensations, thermal sensations, etc. Two or three times he had reconstructed an entire day. He told me: *I have more memories in myself alone than all men have had since the world was a world.* And again: *My dreams are like your vigils.* And again, toward dawn: *My memory, sir, is like a garbage disposal.*"

Borges' narrator also tells us that he suspects Funes,

> ". . . was not very capable of thought. To think is to forget a difference, to generalize, to abstract. In the overly replete world of Funes there were nothing but details, almost contiguous details."

Fifty years after Borges' composition of this unforgettable story, we are confronted by the equivalent of Funes in our own publishing lives. With our comprehensive data bases, our information-packed CDs and our relentless E-mail, we risk overlooking Borges' admonition that "we all live by leaving behind."

If there were no longer publishers to weigh, to select, to edit, we would be like Funes, invalids in our own garbage of memory.

Editor's Note: the author is Executive Director of the Association of American University Presses.

———— <✑ ————

I am a 52-year-old high school teacher on the Micronesian island of Yap. Our school has no toilets for students but we have the most sophisticated computer lab in the Pacific. So far I have managed to remain computer illiterate.

     Margarita Gibbon
     Colonia, Yap

Information technologies, for all the attention they receive, lag far behind the power of the human brain. Researchers estimate that the normal brain has a quadrillion connections between its nerve cells, more than all the phone calls made in the U.S. in the past decade.

    NATIONAL GEOGRAPHIC

MY LAST CLASS:
THE HMONG VS. HIGH TECH

TESTIMONY BY
LEONORA HOLDER

I WILL NEVER FORGET the last class I ever taught. I was scheduled to teach a semester of the much dreaded Remedial English Computer Workshop, along with my four other sections of American Lit., etc. The purpose of this horror was the alleged teaching of basic English to students from all over the world. My colleagues and I were under the impression that we had been doing just that, happily, and successfully, for some time before the damnable blinking bleating buzzing bloody boxes had ever appeared. We were all eminently "computer literate" (wonderful oxymoron), but there are some things polite people don't do, and one is teach from a box. Some creatively-challenged department head had received a grant or something to develop a program, so there we all were, students and faculty, chained to the beasts.

The irony of attempting to teach writing by means of computers to students who, in the main, could not type, did not entirely escape our notice. Nor could we fail to grasp that many of them were illiterate in their first language, in many cases because there had been U.S. backed wars going on in their countries during their formative years, which precluded everything but survival. Those of us with tendencies toward conspiracy theories, saw a greater plot. It seemed to us that these struggling students, most of them from economically underdeveloped and war-torn countries, were being trained for futures doing slave-waged technological piecework on similar machines, with literacy and the learning of English an accidental by-product if it happened at all.

These truths came home for this teacher in a dramatic way as I rounded the corner to the Computer Lab and what would turn out to be the last class I would ever teach. My eyes were immediately assaulted with as clear a vision of Ancient Noble Culture Meets Tech Ugly as I have ever been permitted. On a table next to the wall stood one of the noisy little boxes, hissing, quivering and making loud, "Bleh!" noises. (I couldn't have said it better myself!) Next to it, on the same table, a Hmong man of many winters knelt, with an expression of imperial disgust. He knelt on the table because no one had yet proven to him the superiority of chairs, and he treated the computer like a bad smell because he understood that the important things about multi-cultural teaching, the osmosis of another person's culture and ways of thinking, along with the lessons, was never going to happen when we all relied on megabytes to do the teaching. I understood very clearly at that moment two things: First, that it would not be me who dragged this ancient warrior of a tradition far older than my own into the rubble of mouse-land, and second, that training people for the lowest levels of cyber-slavery was the teaching wave of the future and it was a wave that would not be sailed by me. At the conclusion of that lab, I carefully arranged for substitute instructors and filled out all the class paperwork. Then, after 17 years as a Professor of Language Arts and Literature, I walked out of the lab, out of the college and I never looked back. I hope Thoreau, whose writings I taught for many years, without the interference of a box, would have been proud of me.

———— ————

I recently "retired" from IBM after 27 years. Certainly I have witnessed the rush to nowhere and in fact have been a part of it. Technology, that is leading edge computer technology, is only for the very few. The rest of us, however, are expected to keep up and regrettably end in frustration if we try.

I drive a 1974 car. I do not have a PC in my home (we do pantry inventory manually).

> Bruce Dayley
> Medford, New Jersey

THE MEDIA-FREE FAMILY

TESTIMONY BY
MARY ANN LIESER

MY HUSBAND AND I have lived without television for a long time—the four years of our marriage plus at least five prior years, separately, for each of us. It has been more recently that we've given up radio as well. For over a year now we've lived without voices in our home save those of the real, live people who live here or those of visiting friends.

We talked about it for a long time. I believed ridding our home of radio would be a positive move for us, individually and as a family. And I also resisted the idea. I listened to several hours of public radio news broadcasting morning and evening, and I was reluctant to give that up. "Later," I said. It was autumn of an election year and I wanted to listen to the presidential candidates debate. We put it off time and again. Then one morning over breakfast we decided to drop off at Goodwill that day all the radios we owned. And so we did it.

I felt light and free, as I always do when I give away things I no longer need, clearing out a little more space in my life and my home. I also felt as if I were at sea. How would I know what was going on in the world? How would I keep up with things and stay connected? Those questions—of such importance to me then—seem almost silly now, but it's been a long journey to get from there to here.

65

A week passed before I stopped reaching for the radio in the kitchen each morning to listen to the news before I even started breakfast. The next thing I noticed was how oppressive silence can be. When our daughter was learning to talk, shortly before we gave our radios away, she would point at the radio that sat on our kitchen counter and say, "guy talking." When she noticed it was gone and pointed to the empty space it used to occupy, I told her, "no more guy talking. It's just us talking now," and I wondered how we would fill up all that space of silence that used to be filled with voices from the airwaves.

With nothing to fill the silence but my own voice, I began singing. I had been singing to my almost two-year-old daughter since her birth, but only a song here or there. I could quickly run through all the songs I knew the words to, and then find myself struggling to remember some more. What were those ones we sang around the campfire in sixth grade? How did the second verse of "Go Tell Aunt Rhody" go? I sought out songbooks at our local library and relearned dozens of songs from my childhood, songs that I'd not sung in years, and it was fun. We've learned many new songs together, too, ones I'd never known before. Hymns and songs of praise, old folk tunes, lullabies, silly songs, sentimental ballads—our heritage in music.

I AM NOT a good singer. My husband still winces sometimes when I switch keys in the middle of a melody, but he would be the first to say that my ability to carry a tune has improved a hundredfold. Best of all, I discovered that I love to sing. I sing unselfconsciously all day long and so does my now three-year-old daughter: "Clementine" as we do the breakfast dishes, "Greensleeves" to the baby as I rock him, "Riding in the Buggy Miss Mary Jane" as we sweep, "Down in the Valley" as we tidy up the livingroom, "Old Dan Tucker" as I hang the wash out on the line, "Joyful, Joyful We Adore Thee" as we set off on a walk.

Human voice fills our house, and is richer than any electronically-generated sound could be. My children, now ages one and three, live in a world that is full of music, and they have learned by example that singing is something real people can do for the pure joy of it, not as something we pay entertainers to do for us. And they

Spellcheck did not like my last name, it suggested as substitutes "glum" and "slum." I believe that I would have found "blue" and "plum" less hostile.

Marian Blum
Waban, Massachusetts

In Ray Bradbury's Fahrenheit 451—which was written in the early 1950s, just after televisions and computers first appeared—people relate most intimately with electronic screens and don't like to read. They are happy when firemen burn books.

Cram people "full of noncombustible data," the fire captain explains. "Chock them so damned full of 'facts' they feel stuffed, but absolutely 'brilliant' with information. Then they'll feel they're thinking, they'll get the sense of motion without moving."

67

have learned that it is definitely not something you have to be very good at to feel worthy of doing.

Everyone is worthy of the joy that can come with a spontaneous song, but many people who have electronically-generated voices as a constant presence with which to compare themselves, believe they should possess perfect pitch and a backup orchestra to be worthy of singing. We don't compare ourselves to anything, except maybe the silence, and to us our voices are more beautiful than anything that ever came into our home over the airwaves.

We don't sing or talk all the time, though. The silence that I was always compelled to fill up has become beautiful to me as well. I've become accustomed to being quiet with my own thoughts, and now—especially when I do chores such as washing the dishes, I can be centered and meditative in a way I never could be when I either was listening to talk radio or had voices from it still echoing in my head. This calm and centered way of being in my daily life fits better with the slower pace toward which my family is striving.

Radio has become like so many other things we've given up, or made a conscious decision to live without, from television to a clothes dryer. Our lives become much richer, in more ways than we could have imagined when we were living with what we have since given up, whether it be a convenience, a luxury, or a labor-saving device.

ALTHOUGH I DON'T have the freedom to listen that I had with a radio—freedom to hear over a dozen stations at any hour of the day or night—I have a different and more fulfilling freedom: freedom from having radio voices and noises in my house and in my head. And of course I'm free of the compulsion to turn the radio on to alter my mood or distract myself from my own thoughts. Often when I had a radio I would feel compelled to listen to the news even when I didn't really want to, just to make sure I wasn't missing anything. Now that the choice is removed, I have much more freedom to choose when I want to learn about the news of the day.

AND WHAT ABOUT "missing" things? How do I keep up and stay connected? How do I know what's going on in the world? Now that I've lived this way for over a year, I can

Bradbury's novel no longer seems set in a distant future. Thanks to growth in computer capacity, television and computers are merging into Digital streams of sounds, images, and text that make it possible to become absolutely brilliant with information.

NATIONAL GEOGRAPHIC

At the time of your interview, I was driving across the Blue Mountains. I stopped at my usual garage, a little roadside concern at Bilpin which is a truly peaceful, beautiful spot. The owner seemed shocked that I wasn't in a hurry and that I had time to yarn about the weather, etc. He said most of his customers were city people rushing to and from Sydney who grew extremely impatient if kept waiting for more than a few minutes. . . . Hearing you talk ✏️ safely conclude that those questions that were of such concern to me really aren't issues at all. The news — developments in world or local events — is always there, and I often have a better and deeper understanding of such events when I can spend concentrated and focused time once a week reading about them. I can scan reading material and decide how much time I want to spend reading about any given topic. I have the power to decide, rather than letting National Public Radio choose how much of my mental time will be devoted to Bosnia and how much to Somalia.

Not having a radio actually means I'm better informed about world and local events. Not having a radio means I can better focus my attention on my children playing in the room while I make supper, thus I'm better attuned to their needs and moods. Not having a radio means I have more control over what fills my mind, and it means I'm more comfortable with and accepting of silence. Not having a radio means I've learned more about silence, about the many kinds and textures of quietness that can fill a house. Ultimately, not having a radio means for me that I am more fully human, more involved in forming my own mental landscape and in relating to the other people in my life.

And I can't help but believe that my children are growing up with a much richer childhood for not having electronic media in our home. They too are learning about the textures of quietness. And of course they are doing what children are best at: learning to fill up the silence in their own way. Right now my daughter is making up her own verses to "Here We Go Round the Mulberry Bush" and singing them to her doll, and my son, though he doesn't yet talk, is babbling the tune to "Skip to My Lou" on key.

about the ideas of the Lead Pencil Club on Australian radio after the incident at the garage was a turning point in my life.

Lesley Stoneman
Phegans Bay,
Australia

Whether or not it draws on new scientific research, technology is a branch of moral philosophy, not of science.

PAUL GOODMAN

71

I have long been a fan of handmade books, picture postcards, wind-up toys, and rowboats. My days are spent as a "communications professional" in front of a cathode ray, but my heart's not in it.

Kenn Compton
Mathews,
North Carolina

LETTER PERFECT

TESTIMONY BY
RON TANNER

Blessed be letters—they are the monitors,
they are also the comforters, and they are
the only true heart-talkers.

I. K. MARVEL

VIRTUALLY EVERY middle-class adult in America at one time or another has received a letter, what Goethe called "the beautiful, the most immediate breath of life." Who doesn't want one? Who doesn't know the thrill of tearing open that envelope, its flap gummed with the sender's tongue; withdrawing that piece of paper, redolent of the sender's perfume or hand soap or house odor; and seeing that aged, quaint, tender salutation "Dear . . ." Not even a conversation, not even an intimate face-to-face in the corner booth of your favorite cafe on a rainy afternoon, carries the same charge—because a letter is a possession in a way a conversation is not. Whereas someone might overhear a conversation as it occurs, no one can read the letter in your hands without your knowing. It has the immediacy, the intimacy, of a secret. Signed and sealed. Just for you.

This desire for something so very private, so very personal, accounts for the state of the American mail today. That blizzard of junk we receive daily would not be possible if we did not want, hope, long for a letter from somebody, anybody. We'll open anything that appears to offer, even half-heartedly, a personal address—which is why savvy advertisers insert your name or my name in the appropriate slots of their computer-spawned solicitations.

> "I was flying home to Baltimore from Chicago on a packed holiday flight," a friend writes to me. "Somewhere over Ohio, my nose suddenly began bleeding. Gushing, really—a blood faucet. I don't normally get nosebleeds, so I have no clue why this happened. It was like a god-awful horror flick I had just seen called Children of the Corn, where an evil demon-possessed Satan worshipper uses his mental powers to cause the nose of a God-loving church-goer to suddenly erupt in a fountain of thick red blood, spraying the entire congregation with goo and gore. Blood was literally pouring down my face.
>
> "It dripped all over my shirt as I looked around for something to stop the flow—but there was nothing."

Though historically the letter was a matter of convenience (i.e., the best means of getting out the news), it was never a matter *only* of convenience. Those who prefer the telephone or, worse, E-mail to letter-writing miss the point. E-mail is for the transference of information and nothing more. Utilitarian in the extreme, its messages are usually in the form of unedited one- or two-line bites, almost always in response to a query or a request. It is as spontaneous as a telephone conversation and there lies the problem. A letter is never spontaneous, though it may have the appearance, carry the tone, of spontaneity. (Letters which are no more than spontaneous scribble we call "notes.") A letter is a recollection, an orchestration of thought, often carried by a narrative undertow: there are events to recount, significances to reflect upon, personal inquiry to pursue. What we impart, above all, is not information but, rather, something of ourselves—a view to the mind and heart of the writer.

"I tilted my head back and squeezed my nostrils, but the blood was just getting everywhere," my friend continues. "The man I was sitting next to looked over at me, quite startled, probably wondering if I was having some kind of seizure.

"He pulled out a napkin from somewhere and handed it to me. As he did, some blood got on one of his fingers.

"Eventually, a stewardess came and brought piles of paper towels. When we landed, I had fistfuls of crimson towels.

"After handing me the napkin, the man wiped the blood off his finger. Then, he looked increasingly nervous. I wondered if he feared he might have just contracted AIDS.

"Sure enough, he blurted it out: 'Listen. I'm sorry, but right now I'm not very comfortable. In fact, you just scared the shit out of me.' He paused, taking a deep breath. Because of the cramped flight, we were literally touching. It was a weird forced intimacy—really uncomfortable for me, too. 'Some of your blood got on me. Do I have anything to worry about?'"

Letter-writing (which should be distinguished from note- or card-sending) has always demanded time and art. Time dedicated to getting the sender's story straight, which is to say eloquent, and art to make that eloquence sound "natural," even spontaneous. Consequently, letter writing early on belonged to the elite, the well-educated, the leisured. But eventually, thanks to cheap paper and public education, letter-writing became Everyperson's literature. Many of the first novels written in English were epistolary, like Samuel Richardson's *Pamela*, his housemaid heroine writing, in letter after letter, of her efforts to thwart the lascivious attentions of her employer. Pamela announced to the world that not only might anyone, even the commoner, have eloquence but, with pen in hand, power. Letter-writing helped foment a democratic revolution in America, let us not forget.

Letter writing is still, and I dare say will remain, the most democratic of

75

I am a primary school teacher at Mt. Helena Primary School which is in the Eastern Hills about 60 km east of Perth. My class of 11-year-olds this year are a ✏️ communications. A stamp, paper, a pen. Who can't afford that? In comparison, computers, even phones, cost big bucks, though no doubt computers will be as common as phones one day. Give it time.

Problem is, it takes time to make a letter. Half an hour, at least. Notice I write "make," which is something you don't do when you E-mail a message. You "input data." No sweat, no thought, because it takes no time. Just do it, fingers spidering over the keys, and presto! Yeah, it's fast, it's convenient. And it leeches one's need to write letters, because letters are about the "news" in the broadest sense of the word, not simply the obligatory update—the kids are fine, Jack's taking dance lessons, I still watch "Roseanne"—but a view of the nuances of daily life, of the small events that make one day different from the next. Like my friend's anecdote: "I didn't know what to say," he writes of his neighbor on the airplane. "Of course, I didn't think he had anything to worry about. I had my blood checked during [a recent medical procedure] . . . and I was clean. But it was so strange—plunged into this prickly, unsavory closeness with this other person, having to think about the contexts of my bodily fluids and report the results to a man whose leg was pressed against mine.

"Then again, I really don't know for sure. I haven't had unsafe sex or used needles or tongue-kissed anybody with AIDS, but hell, I don't know. Maybe my dentist gave me syphilis. Maybe I picked up measles in Iowa. What should I say to this man?"

This friend, a computer wizard who's nearly young enough to be my son, has encouraged me to open an E-mail account so that he can "access" me freely, for he and I enjoy each other's conversation. I have refused, however, because I *like* writing letters. My friend is himself a fine writer—which is another reason I will not "communicate" with him via the Fishnet or Hairnet whatever the hell they're calling it. I want to read what he has to think. So he indulges me. His recent letter, like most letters, answers the obligatory questions, How are you? What's new? But, whereas a telephone conversation or E-mail encourages a cursory reply ("I'm fine. Work is a drag, as usual, but I'm getting along . . .) the letter, composed in isolation and comprised of recollection, compels elaboration. How is my friend? Generally speaking, he's fine, all's well. But his nosebleed anecdote—

particularly diverse lot with the common characteristic being that they are all addicted to video games, computer distractions, television inanities, and radio waves.

When it comes to being able to think creatively, they would just have to be the most backward bunch I've taught in the whole of my 26 year teaching career.

Rick McKellar
Western Australia

Reading is a far more demanding and exacting undertaking than anything the new technology has to offer. Reading demands that the imagination be put to work; electronic interaction and its various spinoffs substitute visual and aural images for imagined ones. Reading requires an engagement between reader and text; interaction, which by no means is the centerpiece of his letter—touches upon something we all live with in the late twentieth century, a certain fear, a certain suspicion. AIDS has become a metaphor for the Unknown, for how little we truly control our lives, not to mention the world, in what once seemed an era of Progress and technological mastery. A simple nosebleed can wreak a quiet havoc among us, never mind we have the power and the wherewithal to jet from one country to the other. Or to compute data from one continent to the next.

My friend ends his anecdote: "Definitely nothing to worry about, sir. I just had a doctor's appointment, and there's nothing to worry about, thinking back to the medical . . . [procedure] in the spring. Was that a lie? Should I have told him I had no real idea, though I expected [sic] I was safe? Should I have given him my name and address so he could follow up? Should I have told him if I was gay, if I was a heroin user, or if I ever had been?

"I don't know why, but that incident has really stuck with me. Especially the look on the guy's face when he saw blood on his finger—genuine fear, a fear caused by me. A fear caused by AIDS. But, to me, principally a fear caused by the fact that I was a strange, unknown person, and that my bodily fluid had the potential to kill another man."

Could my friend have, *would* he have, sent me this on E-mail? Would a conversation, etherized in immediacy and corroded by memory, have had the same impact as these, his own words—as bold as bloodstains?

The E-Mail version: "I had a nosebleed on the plane ride home. The damnedest thing. Blood splattered on the guy sitting next to me. He thought I had AIDS maybe. Did I? Do I? I told him not to worry. But it makes me wonder. Scary stuff."

The problem with telephone conversations is that they are subject to interruption—which is why, whenever I have important business to conduct on the phone, an interview with a prospective employer, for instance, or divorce arrangements with my soon-to-be ex-wife, I *write down* what I have to say. The potential to lose one's thoughts in conversation, to sidetrack, to stall, to dead-end, is tremendous. Many times I have composed a letter to a friend, a lover, or a family member to make clear what I did and did not say in a recent telephone conversation.

however alluring, is game-playing. Reading at its most demanding is, purely and simply, hard, which is precisely why we are so eager to be done with it.

JONATHAN YARDLEY
WASHINGTON POST

I am the last administrator at the University of Michigan who is totally computerless. Here's a quote for you from John Pendelton Kennedy, "Industry can only be found where artificial wants have crept in and have acquired the character of necessities."

Beni
Ann Arbor, Michigan

I want to join the club. Credentials? I rejected the offer of a fax for Christmas, opting for a bottle of ink and a quill pen . . . I hope I qualify.

Judith Crist
Woodstock, New York

79

Obviously this is what makes conversation exciting, whether on the phone or E-mail or via the next great technological breakthrough (cerebral implants?): its spontaneity, its digressiveness, its refusal to belong to anyone in particular. Certainly a letter, by comparison, is intimidating, especially in this litigious age. It is a signed document, after all. More specifically, a letter is a monologue. Its value, its significance, is that nowhere else in the letter-writer's life will these events have been so well-framed, here in this concrete form, shared with a particular person, for a particular reason.

We can obfuscate conversation with subsequent remarks and a faulty memory, and E-mail we can dismiss lightly not only because it's off-the-top-of-the-head but because its brevity automatically excuses us from committing ourselves too fully. But a letter we cannot recall or retract because, even after we renounce the last one we sent, it remains in the recipient's possession. Someone I know in academe has been unable to complete her Ph.D. dissertation on a famous playwright because the writer's family refuses to release his letters to public scrutiny. They know that too much can be made of the written word.

My mother, of an old southern tradition, has always enjoyed writing letters and, remarkably, her many friends, now removed to many parts of America, write back. I say "remarkably" because my many friends, also removed to many parts of America, seldom write, if they write at all. We're all so busy, aren't we? Life eats us up. It's a generational problem which clearly underscores what we are losing: the time to write. Fifty years ago Emily Post observed: "The art of general letter writing in the present day is shrinking to such an extent that the letter threatens to become a telegram, a telephone message, or just a postcard." Or an E-mail message.

Let me retract that. The problem has to do with time, as I suggested, but it's not a *lack* of time that prevents us from writing, it's a lack of understanding, which is a product of the time in which we live. Those who don't like letters would like them better if they saw letters not as an obligation—especially not a means of freighting information—but as an indulgence, an opportunity to spend some time with their thoughts and to share something of themselves with their friends. The letter was, after all, the precursor of the essay, that leisure enterprise of

ALWAYS BRING A PENCIL

There will not be a test.
It does not have to be
a Number 2 pencil.

But there will be certain things:
the quiet flush of waves,
ripe scent of fish,
smooth ripple of the wind's second
*   name,*
which prefer to be written about
in pencil.

It gives them more room
to move around.

NAOMI SHIHAB NYE

81

I believe I would qualify for membership in the Lead Pencil Club as I am undoubtedly one of the very few editors at high tech CIA that still uses a #2 pencil instead of a computer.

Paul E. Arnold
Manassas, Virginia

intellectual meandering (Montaigne tried to emulate the eloquence and simplicity of Seneca's letters, you remember). Most people who don't write letters feel intimidated by the prospect of penning one—it seems to them an extraordinary feat—in part because they aren't in the habit of writing, but in greater part because they feel compelled to be scintillating, witty, wonderful. This misapprehension grows not so much from the letter-writing tradition as from our entertainment-obsessed culture. Television in general and television talkshows in particular encourage us to believe that when the spotlight illuminates us, we damned well better have something stunning, if not shocking, to share.

Consequently, those who would write letters, don't because they feel too much pressure to perform. They are never up to the task, their lives seem so boring—at least compared to the gore they see on TV—and it seems they never have enough stuff saved up to write about anyway. The one exception to this shortfall is the Christmas or New Years letter, which a sizeable number of Americans exchange annually, usually a multi-paged epistle of anecdote, gossip, and family chronicle. Ironically, anyone who has written such a letter realizes that there is too much to say, that too many events and explanations must be compressed, truncated, or expunged in order to accommodate this once-a-year reflection.

The greater irony is that letter-writing has returned to its original province, an indulgence of the privileged few, except that the privilege nowadays is simply a matter of knowing the pleasure of writing letters. It's not as though no one has the time—turn off the TV for an hour—nor is it that one needs great erudition or literary acumen to fill a page or two—children sometimes write wonderful letters—nor is it that there is nothing to say—we are human, after all; we have plenty to write about. And yet so many believe that letter-writing is beyond them, never mind that tomorrow most of us will go to the mailbox, as we do nearly every day, in anticipation of receiving something private, something personal, something addressed just to us.

Cyberspace is private space, not public space: It is a set of silicon bubbles, not a town square with a soapbox. In cyberspace people are alone, pretending to be in public. In real public space they are physically together, and if they may sometimes pretend to be alone, that doesn't change the basic fact that they must, in a profound and serious way, acknowledge each other. There are social graces—nods, handshakes, not interrupting while another is talking, courteous farewells—that are a normal part of even the most casual human interaction.

In cyberspace such things barely exist.

PAUL GOLDBERGER

83

Abolishment of Childhood?

TESTIMONY BY
MARY CLAGETT SMITH

As an early childhood program director working among young children, I am struck daily with the increasing blight of our techno-life invading and replacing the here-and-now world children best learn from. The "original" tactile, responsive world of sand, mud, water, grass, and Teddy Bears is rapidly being replaced by screen simulation. Children are learning the basics second-hand—from two screens, the TV and the PC. Infants grow up beside a family member who babbles (or shrieks) in the living room corner all day and evening. Research shows that the four hours per day (national average) impressionable children watch TV gives them more contact with sitcom and cartoon characters' values than their own parents'. Children learn the most from those they have the greatest access to. Beavis and Butthead? The Power Rangers?

Who is raising our country's children? The marketing industry? Where are the parents? They both arrive home limp and exhausted from breathless days shouldering cellular phones while faxing info, E-mailing memos and clutching for instant Internet data. Weary parent and dazed child collapse together every evening to gaze blankly at yet another screen. Are we fast losing touch with enough close, slow, warm doing/pondering time for children? *With* children?

We better bring back the real, hands-on childhood so children can know at a base level what the world really is. They need to feel, smell, see, hear and wonder about life, not just watch a simulated one. A new set of 3 Rs is in order: **R**eal, first-hand processes and experiences like digging up earth-grown carrots, or making a "fort" out of old sheets, or raking and piling brown, crunchy leaves to nest in; **R**aw materials to mess around with—wood to saw and nail, mud to shovel and spill, cloth to cut and stitch, flour for kneading and baking their own bread, and crayons, scissors and paste to create stuff; and **R**esponsive grown-ups who slow down to listen, sing songs, stir soup, laugh at clouds, counsel, tell stories and take hand-in-hand walks in the foothills or along riverbanks. If we adults stop to smell the flowers and invite children to sniff too, we'll restore our senses, revive our humanness, and recognize our children's loveliness.

85

We all know that E-mail is supposed to facilitate transglobal friendship and maybe even world harmony. But might it also spawn irritation, anger and shame? Already dozens, if not hundreds, of academics have become notorious for the stupid, insensitive, or arrogant comments they've posted on the now ubiquitous discussion lists set up for every topic from Chicano Literature to Diplomatic History

LINGUA FRANCA

Soon computers will be able to do anything for us and we will be mindless cabbages in a world run by our creations.

Jess Hester, Student
Baypoint Middle School, Florida

Chip Bok
Akron Beacon Journal

ASIA, ASIA!

EVA SCHAWOHL

THE PICTURE—a Buddhist shrine in a small Sri Lankan town, the entire chamber is alive with the many images of the Buddha. Buddha at rest, in meditation, in starvation, in Enlightenment. Heavy whirls of incense smoke cloud the multi-colored chamber and the intent worshippers clasp their hands above their heads, chanting. Outside, a monk, as fat and jolly as the Buddha himself, steps out of a chauffeured Pajero jeep. Western civilization has come to Asia.

Another scene. A village in Bangladesh, with no water save for one deep tube well, no school building. Its inhabitants are farmers who pay with their sweat to landowners. Yet on the roof of a thatched hut is a television antenna and yes, there is a television. To their delight, the farmers can now watch soap operas. Welcome to the wonderful world of Western technology brought into the heart of Asia, courtesy of J. R. Ewing.

Does Asia need this kind of technology? It clashes with Asian culture and Asian needs. Instead of providing a school to that Bengali village, the aid agency provided a television—"for educational purposes." The farmers cannot read or write, they keep their wives in purdah and have at least twelve children so to enhance the quality of their lives they now have television. Educational purposes, naturally! Perhaps by providing television, someone out there in the West is hoping the farmers will be too distracted to think of copulation.

In the last three years, Asia has been experiencing an incredible boom time, comparable perhaps to the California Gold Rush. In a stale city like Colombo, money became the watch word—a battle of haves against the ones who have even more. Shopping malls sprang out of the ground, BMWs became common road accessories, computers were installed everywhere from hospitals and government buildings to private homes. The biggest leap forward came with the introduction of the cellular telephone. First a company called Celtel, followed closely by Call Link, and nowadays they are into Motorola. Ironically, it is still a game of nerves trying to call anywhere out of Colombo. It is technology without a doubt but it is going up a one way street. A majority of the population cannot afford a Celtel but they would appreciate a decent telephone service. This does not really concern the haves who, after all, do not have to rely on the local Telecom anyway. It is a matter of affordability.

What it boils down to is misinformation. Granted Asia does need to have access to Western technology in order to someday become competitive in their own right. However it is the wrong kind of technology that is getting the best foothold. In the past year, the gap between the rich and the poor has become unbridgeable, with the capital cities swallowing all the Western trinkets. Individuals benefit, not the masses.

In a country like Bangladesh it is a direct fault of Western donors, too much doing and no planning. They have given technology which the country is simply too illiterate to understand. One can almost picture the well-meaning ladies of the country club sitting around their pot of tea and discussing the year's donations—"But it is simply terrible! To think all those poor people in Bangladesh do not have computers! We have had them for simply ages and it is just so cruel to them that they should be left behind in such a thoughtless way! Let's donate twenty thousand with the next flood relief container. It'll make them so happy!" Bangladesh is drowning in computers that very few have the knowledge to use and televisions which can now be hooked onto the Star TV Network, via satellite.

The cycle of too much technology,

A bureaucrat armed with a computer is the unacknowledged legislator of our age, and a terrible burden to bear. We cannot dismiss the possibility that, if Adolf Eichmann had been able to say that it was not he but a battery of computers that directed the Jews to the appropriate crematoria, he might never have been asked to answer for his actions.

NEIL POSTMAN

89

too little education, is very well established now throughout Asia. It is not possible to stop—computers, cordless phones and expensive cars have become symbols of prestige, of goals to strive for, even for the poor. They will carry on wanting them long after we have discarded them as yesterday's hat. Yet it can be prevented from getting worse. The answer lies in education. An educated population that can decide whether it wants a computer or a BMW is one that can also decide to better itself through other means. At the moment, that population is not choosing for itself but being force fed. The rich buy not out of need but out of greed and then present this image to the poor who want it too. Why does the monk for example have a Pajero? In this car, he is a symbol of power—and power is what he wants over the worshippers. He can afford it so he is better and to be followed unquestionably. It is too late to undo such a grievous mistake but it can be prevented from getting worse—a lead pencil can play a tremendous role in bettering the lives of Asia's people, simply and efficiently, providing literacy and the very freedom the West takes for granted—the freedom of free thinking.

YOUNG CYBER ADDICTS

TESTIMONY BY
AMY WU

THE PHENOMENA of sending letters to all of my high school friends with the touch of a button, and joining the "rec.music.Dylan" for diehard Dylan fans, transformed me into a shameless cyber addict in my freshman year.

Like the hundreds of other bleary-eyed addicts I made my nightly trek to the computer lab, where a queue-shaped waiting line was already formed, my fingers itching to touch the keyboard and my mind already set on chatting with my on-line boyfriend R2D2. For weeks I forfeited sunshine for a fluorescent terminal. A whole new world was opening up before me until my "A" average in anthropology drifted to a mediocre "B."

The Internet is becoming young America's latest addiction, especially on college campuses. The "Just say No" to sex, drugs, and alcohol may soon pertain to E-mail and surfing the information superhighway. Soon CA (Cyber addicts Anonymous) may be added to AA. Blame it on the free E-mail accounts and easy access most colleges offer.

A cyber addict is as easy to distinguish as a swaggering alcoholic. They sit before a screen for hours laughing, talking and smiling at a screen, hopelessly lost in their own world. They proudly tell you that yes they do spend four hours a day in cyberspace, that yes they have a fruitful social life where they chat with friends with names like KillBarney, that yes they procrastinate on major papers so they can keep up with their E-mail correspondence.

When asked what they would do if the school took away their most prized possession they gasp and turn pale with the possibility. "If I didn't have access I'd have to get a life," a junior, said with a nervous laugh.

For others it would be more of an inconvenience than a total loss. "I wouldn't freak, some people would just freak," a freshman said. Others say that they would just buy a modem. All they have to do is log on to the school's system, still free of cost.

The temptation of entering cyberspace is great for many young people. It's a cheap and quick alternative to snail mail (the kind with a stamp). The "Talk" channel bears an uncanny resemblance to the telephone. Logging onto the Net allows you to chat with as many as thirty people at the same time.

There are hundreds of newsgroups where the latest movies can be debated, where the psychology of body art can be dissected and where Camille Paglia and Rush Limbaugh can be bashed.

Instant friends from as far away as Australia and Africa are made through the "soc.penpals." Love at first byte is even a possibility. My girlfriend and a cyberfriend went out to a cafe after meeting each other online. Unfortunately conversation over cappucino didn't make them compatible, and they never E-mailed each other again.

The Internet is a channel for the curious. With the click of a mouse the Dead Sea scrolls can be viewed, the President's health care plan will appear, and letters to editors or to cyberzines can be written.

For a generation accustomed to fast music, fast food, and quick results, the Internet is a perfect match—it's easy, it's fast, it's fun, it's free. It is also addictive and as dangerous as it is educational.

There are stories of young people who have disappeared into the computer, who become so addicted that they cybersurf for nine or ten hours and continue into the night. There are always one or two of these hopeless addicts in the lab. They have glazed expressions on their faces and

My daughter, Laura, who is only 3 years old, has drawn this which I believe is how one looks after staring into a TV for one hour. . . . Certainly we must spread the idea that increasing technology leads to depersonalization, loneliness, lack of humor, and possibly, drive-by shootings.

Michael Clay
Laguna Hills,
California

93

if you wave to them they think you're a figment of virtual reality instead of reality.

The Dead Sea scrolls aren't addictive, but mudding—a Dungeons and Dragons type game—and checking how many E-mail messages you've received is. One young woman tearfully told me how she became addicted.

"Oh I was bad," she said. "I wouldn't count the hours I would just be there." She and her channel friends would chat about everything from Nine Inch Nails's newest album to rumors of Kurt Cobain's ghost. Her addiction reminded me of my addicted roommate who begged me to hold her computer card for a week after she had done poorly on a mid-term because she spent the previous night on the Internet. Needless to say the week didn't last.

Another young woman, whose grades plummeted from 3.3 to 2.5 after getting hooked on the Internet tried to explain. "I didn't stop studying," she said, "you just like get addicted to it. You're so into the conversation you don't want to get off and study, you just study less." It's a typical response from a cyber addict.

These young people are drawn into a world where they are connected to the world but sadly disconnected to their environment. Many have lost friends and a social life which includes going to the movies or out for pizza. Some haven't talked on the phone and written a letter for a year. Many have given up school activities, student government, and sports.

Earlier this year my suitemate's boyfriend even called my room and told me to tell her to get off the computer so he could talk to her. "Get off the computer!" I screamed. "Don't bother me!" she shot back. The boyfriend was despondent and inconsolable and threatened to throw her computer out the window.

Many addicts lock themselves in coffin-styled dorm rooms and spend sunny days sitting under a fluorescent light staring at a screen for hours until that glazed expression is achieved.

If only colleges charged Internet use per hour, if only accounts were given out on the basis of need, if only hours were limited, then maybe horror stories about young people who have failed classes because of too much E-mail, who have disappeared in the mudding world, who haven't spoken to a human being in a week, who haven't felt the sun in days, and are proud of all the above, wouldn't exist.

As seemingly great as free access on

I think Internet is about the most antisocial thing I ever heard of, except maybe for guns.

Roberta Pliner
New York City

When the virtual becomes the real, then the computer becomes our window onto the world, and the bandwidth the door by which our friends come to visit. We'll naturally do everything (buy anything) to keep that window open. In time, we may come to fear its closing like death itself.

MARK SLOUKA

95

In 1942, my pal and I spent long Aleutian days inventing post-war products that would make us rich. Our best was a laxative named HASTE, with the surefire slogan, HASTE MAKES WASTE!

Willard Olney
Hesperia, California

the information superhighway may seem to prospective students and parents, the dangers and damage it can cause outweigh the positives.

"I know some people who had to cut down because they developed carpal tunnel syndrome," one young woman said. She stared at her own atrophying wrists, the result of one too many hours chatting with the "Mystery Theater 3000," smiled sweetly and said she had to finish E-mailing Darth Vader, her new online boyfriend. It was just another sad story from a young cyber addict.

— ✐ —

A friend of mine, a distinguished explorer who spent a couple of years among the natives of the upper Amazon, once attempted a forced march through the jungle. The party made extraordinary speed for the first two days, but on the third morning, when it was time to start, my friend found all the natives sitting on their haunches, looking very solemn and making no preparation to leave.

"They are waiting," the chief explained to my friend.

"They cannot move farther until their souls have caught up with their bodies."

STORY RELATED BY EDIE ZIMMER
KAYSVILLE, VERMONT

I am an Episcopal priest who is very interested in the interaction between technology and theology. Obviously, your club has hit a nerve deep in the soul of humankind. I believe that we are indeed overwhelmed with gadgetry and underwhelmed with relationships.

Rev. Gary C. Schindler
Springville, New York

WHEN EXECUTIVES SHOULD JUST SAY NO

**TESTIMONY BY
MADELEINE BEGUN KANE**

I AM A LAWYER who works in a modern office. I can transport reams of documents across continents in an instant or close a multimillion-dollar deal in a flash. I know I should feel grateful. So why do I feel enslaved?

In the pre-fax, pre-computer, pre-Federal Express days—about 15 years ago—life was less complicated, expectations more reasonable. Deals were done at a temperate pace. The corporate merger that we now race through in weeks was cautiously closed in months. Another day or two rarely mattered. Briefs were brief, and an automatic collating, self-stapling machine didn't spew out documents by the dozens. Yes, we were limited by our tools. But that wasn't necessarily bad.

In those pre-high tech days, we were content with carbon copies. We wrote concisely to avoid typing through the night. And we wrote it right the first time. Or at least, close enough. Starting over was wasteful, so we put up with a few smudged corrections and an occasional crossed-out phrase. And we managed to wait for the mail.

Well, we needn't wait any longer. Computers and facsimile machines have taken care of that. And they do come in handy when time is running short. But the problem is that as soon as a technological advance is made, it becomes not a frill but an imperative.

Progress should bring us ease and convenience. Instead it dispenses pressure. We can cut complex deals on fast-forward, so we do just that. Within hours of a handshake we produce 20 sets of 60 word-processed pages memorializing the transaction. We process words faster than we can read them. Then comes the battle of the faxes. Revision requests surge through phone lines—often at the end of a long day as we're bolting out the door. Yes, there are times when every second counts. Restraining orders and Federal deadlines come to mind. But what about all those other times when faxes fly simply because a facsimile machine is there? Or because someone, for no reason other than ego, demands it?

Jumping through computerized hoops may seem impressive. But sometimes it only encourages people to raise those hoops even higher.

So the next time you're tempted to send paper soaring at the speed of sound, examine your priorities. If you're not calling the shots, you may have no choice. But there are times when executives can just say "No." Does that 7 P.M. message really merit an immediate response? Is keeping staff hooked to computers all night a sensible use of personnel? And perhaps most important: Am I using these machines—or are they using me?

SYLVIA

Let's say you're an up-and-coming CEO—young, techno-savvy, wired. You've put PCs on every desktop in your small but fast-growing company. Everyone's using E-Mail. Your engineers on the West Coast zap notes to your marketers back East. Bills get paid faster, customers are better served. You're even doing business on the Internet. But . . . those whizzy computers cost a lot more than you expected. Employees spend way too much time fussing with balky hardware and software. And sometimes when you walk the halls, you hear strange booms and screams coming from behind closed doors. That's the sound of your employees hard at work, playing Doom.

Sometimes you wonder whether you're getting that much bang for your technologic buck. Techies are familiar with the problem. They know it as the infamous "productivity paradox." American companies have spent nearly $1 trillion on fancy computer systems over the last decade—with almost no gain in productivity. Apostles of the Information Age have tried to explain away that awkward fact as a "metrics" problem. How can you accurately measure output, they ask, in an information-based economy? The bad news, most economists now agree, is that the productivity paradox is no statistical fluke. It's real.

NEWSWEEK

THE MYTH OF COMPUTERS IN THE CLASSROOM

DAVID GELERNTER

OVER THE LAST DECADE an estimated $2 billion has been spent on more than 2 million computers for America's classrooms. That's not surprising. We constantly hear from Washington that the schools are in trouble and that computers are a godsend. Within the education establishment, in poor as well as rich schools, the machines are awaited with nearly religious awe. An inner-city principal bragged to a teacher friend of mine recently that his school "has a computer in every classroom . . . despite being in a bad neighborhood!"

Computers should be in the schools. They have the potential to accomplish great things. With the right software, they could help make science tangible or teach neglected topics like art and music. They could help students form a concrete idea of society by displaying on-screen a version of the city in which they live—a picture that tracks real life moment by moment.

In practice, however, computers make our worst educational nightmares come true. While we bemoan the decline of literacy, computers discount words in favor of pictures and pictures in favor of video. While we fret about the decreasing cogency of public debate, computers dismiss linear argument and promote fast, shallow romps across the information landscape. While we worry about basic skills, we allow into the classroom software that will do a student's arithmetic or correct his spelling.

Take multimedia. The idea of multimedia is to combine text, sound and pictures in a single package that you browse on screen. You don't just *read* Shakespeare; you watch actors performing, listen to songs, view Elizabethan buildings. What's wrong with that? By offering children candy-coated books, multimedia is guaranteed to sour them on unsweetened reading. It makes the printed page look even more boring than it used to look. Sure, books will be available in the classroom, too—but they'll have all the appeal of a dusty piano to a teen who has a Walkman handy.

So what if the little nippers don't read? If they're watching Olivier instead, what do they lose? The text, the written word along with all of its attendant pleasures. Besides, a book is more portable than a computer, has a higher-resolution display, can be written on and dog-eared and is comparatively dirt cheap

Hypermedia, multimedia's comrade in the struggle for a brave new classroom, is just as troubling. It's a way of presenting documents on screen without imposing a linear start-to-finish order. Disembodied paragraphs are linked by theme; after reading one about the First World War, for example, you might be able to choose another about the technology of battleships, or the life of Woodrow Wilson, or hemlines in the '20s. This is another cute idea that is good in minor ways and terrible in major ones. Teaching children to understand the orderly unfolding of a plot or a logical argument is a crucial part of education. Authors don't merely agglomerate paragraphs; they work hard to make the narrative read a certain way, prove a particular point. To turn a book or a document into hypertext is to invite readers to ignore exactly what counts—the story.

The real problem, again, is the accentuation of already bad habits. Dynamiting documents into disjointed paragraphs is one more expression of the sorry fact that sustained argument is not our style. If you're a newspaper or magazine editor and your readership is dwindling, what's the solution? Shorter pieces. If you're a politician and you want to get elected, what do you need? Tasty sound bites. Logical presentation be damned.

Another software species, "allow me" programs, is not much better. These programs correct spelling and, by applying canned grammatical and stylistic rules, fix prose. In terms of promoting basic skills, though, they have all the virtues of a pocket calculator.

In Kentucky, as *The Wall Street Journal* recently reported, students in grades K–3 are mixed together regardless of age in a relaxed environment. It works great, the *Journal* says. Yes, scores on computation tests have dropped 10 percent at one school, but not to worry: "Drilling addition and subtraction in an age of calculators is a waste of time," the principal reassures us. Meanwhile, a Japanese educator informs University of Wisconsin mathematician Richard Akey that in his country, "calculators are not used in elementary or junior high school because the primary emphasis is on helping students develop their mental abilities." No wonder Japanese kids blow the pants off American kids in math. Do we really think "drilling addition and subtraction in an age of calculators is a waste of time"? If we do, then "drilling reading in an age of multimedia is a waste of time" can't be far behind.

Prose-correcting programs are also a little ghoulish, like asking a computer for tips on improving your personality. On the other hand, I ran this article through a

I claim that this bookless library is a dream, a hallucination of on-line addicts; network neophytes, and library-automation insiders ... Instead, I suspect computers will deviously chew away at libraries from the inside. They'll eat up book budgets and require librarians that are more comfortable with computers than with children and scholars. Libraries will become adept at supplying the public with fast, low-quality information.

The result won't be a library without books — it'll be a library without value.

CLIFFORD STOLL

I am a retired professor . . . I suspect I became a real technophobe prior to my retirement when I began to get term papers from my students done on word processors — Oh, so perishingly

103

spell-checker, so how can I ban the use of such programs in schools? Because to misspell is human; to have no idea of correct spelling is to be semiliterate.

There's no denying that computers have the potential to perform inspiring feats in the classroom. If we are ever to see that potential realized, however, we ought to agree on three conditions. First, there should be a completely new crop of children's software. Most of today's offerings show no imagination. There are hundreds of similar reading and geography and arithmetic programs, but almost nothing on electricity or physics or architecture. Also, they abuse the technical capacities of new media to glitz up old forms instead of creating new ones. Why not build a time-travel program that gives kids a feel for how history is structured by zooming you backward? A spectrum program that lets users twirl a frequency knob to see what happens?

Second, computers should be used only during recess or relaxation periods. Treat them as fillips, not as surrogate teachers. When I was in school in the '60s, we all loved educational films. When we saw a movie in class, everybody won: teachers didn't have to teach, and pupils didn't have to learn. I suspect that classroom computers are popular today for the same reasons.

Most important, educators should learn what parents and most teachers already know: you cannot teach a child anything unless you look him in the face. We should not forget what computers are. Like books—better in some ways, worse in others—they are devices that help children mobilize their own resources and learn for themselves. The computer's potential to do good is modestly greater than a book's in some areas. Its potential to do harm is vastly greater, across the board.

Editor's note: the author is professor of computer science at Yale.

neat and clean!—but often adding up in content to nothing. I also have several grandchildren who occasionally write me letters from the information superhighway. Somehow, something gets lost there.

Charles Todd
Vero Beach, Florida

Intelligence experts know that there are probably 10 crucial electronic nodes in our society, which if struck could literally shut America down. A terrorist could shut down the Federal Reserve computers if he attacked those computers with an appropriate virus, and spread financial chaos.

ALVIN TOFFLER

BOOKLOVE

TESTIMONY BY
JEROME STERN

I HAVE JUST COME FROM AN EXHIBITION THAT TOLD ME THAT BOOKS WILL BE replaced by electronic libraries, talking videos, inter-active computers, CD-ROMs with thousands of volumes, gigabytes of memory dancing on pixillated screens at which we will blearily stare into eternity.

AND SO, IN THE FACE OF THE FUTURE, I MUST SING THE SONG OF THE BOOK, nothing more voluptuous do I know than sitting with bright pictures, fat upon my lap, and turning glossy pages of giraffes and Gauguins, penguins and pyramids. I love wide atlases delineating the rise and fall of empires, the trade routes from Kashgar to Samarkand.

I LOVE HEAVY DICTIONARIES, THEIR TINY PICTURES, COMPLICATED COLUMNS, minute definitions of incarnative and laniary, hagboat and fopdoodle.

I LOVE THE TEXTURE OF PAGES, THE HIGHGLOSS SLICKNESS OF MAGAZINES AS slippery as oiled eels, the soft nubble of old books, delicate india paper, so thin my hands tremble trying to turn the fluttering dry leaves, and the yellow cheap, coarse paper of mystery novels so gripping that I don't care that the plane circles Atlanta forever, because it is a full moon and I am stalking in the Arizona desert a malevolent shape-shifter.

I LOVE THE FEEL OF INK ON THE PAPER, THE SHINY VARNISHES, THE SILKY lacquers, the satiny mattes.

I LOVE THE PRESS OF LETTERS IN THICK PAPER, THE ROUGHNESS SIZZLES MY fingers with centuries of craft embedded in pulped old rags, my hands caress the leather of old bindings crumbling like ancient gentlemen.

THE BOOKS I HOLD FOR THEIR HEFT, TO RIFF THEIR PAGES, TO SMELL THEIR smoky dustiness, the rise of time in my nostrils.

I LOVE BOOKSTORES, A PERFECT MADNESS OF OPPORTUNITY, A LAVISH FEAST eaten by walking up aisles, and as fast as my hand reaches out, I reveal books' intimate innards, a doleful engraving of Charlotte Corday who murdered Marat, a drawing of the 1914 T-head Stutz Bearcat whose owners shouted at rivals, "there never was a car worser than the Mercer."

I SING THESE PLEASURES OF WHITE PAPER AND BLACK INK, OF THE SMALL JAB OF the hard cover corner at the edge of my diaphragm, of the look of type, of the flip of a page, the sinful abandon of the turned down corner, the reckless possessiveness of my marginal scrawl, the cover picture as much a part of the book as the contents itself, like Holden Caulfield his red cap turned backwards, staring away from us, at what we all thought we should become.

AND I ALSO LOVE THOSE GREAT FAT BIBLES EVANGELISTS WAVE LIKE OTTER pelts, the long graying sets of unreadable authors, the tall books of babyhood enthusiastically crayoned, the embossed covers of adolescence, the tiny poetry anthologies you could slip in your pocket, and the yellowing cookbooks of recipes for glace blanche dupont and Argentine mocha toast, their stains and spots souvenirs of long evenings full of love and argument, and the talk, like as not, of books, books, books.

———— ✏ ————

What is the InfoSupWay? Well, once the sludgy copper cables that make up the phone and cable systems are replaced with brisk and nimble fiber-optic stands, a firehouse of data will stream into your house until you are knee-deep in useful information. You won't just order a pizza from your InfoSupWay machine—you'll track the delivery person using the Global Positioning System satellite network while your machine simultaneously requests the North American Mean Delivery Interval from the Library of Congress, so you can see how long it would take, and tip accordingly.

The pizza's cold? No problem! In the modern world, your pizza will still be cold, but now you'll have the tools to know why. Call up a databank in Helsinki for a full statistical abstract of heat loss by tomato-based circular objects. Cross-reference with the Delivery Interval data and satellite information. Ah hah! Empirical proof the pizza should have arrived hot! You can now shame the manager into giving you a free pizza while you simultaneously ruin his credit rating, and have the driver deported. (This option will be available only on upper tier packages.)

What a wonderful world it will be. Except—it's nonsense.

JAMES LILEKS
SPRINGFIELD REPUBLICAN

Why *The Pocahontas Times* Does Not Have a Fax

TESTIMONY BY
RUSSELL D. JESSEE

THE POCAHONTAS TIMES has a long tradition of being behind the times. From being one of the last hand-set commercial newspapers in the United States to being one of the last businesses in Marlinton [W.Va.] to get touch-tone phones, we have been slow to embrace new technology, and for the most part we are proud of our retro tendencies.

The only exception is that we were the first newspaper in the state to handle all our typesetting and subscriptions with computers. After the Flood of 1985 we had to get back in business and in short order. Computers allowed us immediately to begin printing editorial and advertising copy, and to address labels.

But even with all our computers, Evelyn Withers, our one-woman billing department, still uses notecards and does her math by hand. Old ways die hard around here if they die at all.

Of course, we don't yet have a fax machine. We're still trying to get used to the hold button on our telephones.

While I'm sure plenty of people think our having a fax would be a boon when they need to get something to us five minutes before our Tuesday noon deadline, I have several reasons why I think a fax won't be any help at all.

First, one of the things people most commonly want to fax here are camera-ready ads. Well, folks, so far fax transmissions are not camera-ready. Even if we can reset the type, the artwork is going to look horrible.

Secondly, we already get too many junk press releases in the mail from groups that have no connection to Pocahontas County. What good would it be for us to have a fax if it was always jammed up receiving junk faxes?

Lastly, we're a weekly paper serving a rural county. A rural county's relaxed pace of life, patience and even stoicism are reflected in having a leisurely weekly newspaper as opposed to an obsessive-compulsive daily. With a fax machine, we would give up a little more of our peace-of-mind.

I'm sure we would help some people sometimes if we went ahead and got a fax. Nick wouldn't have to call me every Tuesday with the Elkins Theater schedule, and out-of-town funeral homes could fax obituaries.

Still, I can't help but believe a fax machine would just create a whole other set of aggravations. For now, let's just take it easy, plan ahead and—I know I'm asking a lot—trust the mail.

USER-FRIENDLINESS: BOOK VS. DISK

TESTIMONY BY
STEPHEN MANES

Material World: A Global Family Portrait (Sierra Club Books) is a fascinating volume of photographs, essays and statistics about the way people live around the globe. Conceived by the photographer Peter Menzel, the book portrays the worldly goods and daily life of 30 representative families, along with statistics and commentaries about the countries they inhabit.

Material World from Starpress Multimedia, is also a CD-ROM that, according to the book's jacket flap, "brings even greater life to the subject with spectacular video, breathtaking photography, and stereo sound." If these claims were true, books in general, and this book in particular, would seem to be endangered species. A comparison between bound volume and CD-ROM should help reveal whether reports of the death of the book are greatly exaggerated.

THE CD-ROM requires a 25 megahertz 386DX or faster IBM-compatible computer (486 is recommended), Microsoft Windows 3.1 or higher, DOS 5.0 or higher, four megabytes of random access memory (eight is recommended), VGA/ SVGA or better display quality with 640 by 480 dots and 256 colors, a

111

mouse, a hard drive, a CD-ROM drive with at least 150 kilobit/second transfer rate and a Sound-Blaster-compatible, Sound Blaster Pro, or Roland sound card. (Macintosh requirements differ.) The book requires only a source of light.

Installing the CD-ROM is an interactive process requiring minutes of pointing, clicking, and waiting in front of a keyboard and computer screen. The book can be installed instantaneously in one's hands or lap indoors or out.

On a 17-inch monitor, the CD-ROM can display 50 dots per inch and 256 colors at a time. On pages slightly smaller than a typical 17-inch screen, the book displays millions of colors at a resolution of more than 130 dots per inch. The book's photos are therefore immensely sharper and richer than their electronic counterparts.

Details that leap out in ink on paper are unrecognizable in phosphor under glass. The quality of light, almost palpable on the page, is absent on the screen. And the book's double-page "big pictures" of families with their possessions dwarf all but the biggest computer displays.

The book is organized geographically. A standard information-retrieval device known as a table of contents provides quick access, as does a stunning map based on satellite photographs of the Earth. Moving from the table to the actual contents is fast and easy even though some of the pages lack numbers. Digital placeholders commonly known as "fingers" are not supplied, but using those typically at hand lets you compare two sections with ease.

Navigating through the CD-ROM can be infuriating. The disk is organized along four pathways: questionnaire, countries, lifestyles and families. Sometimes you can jump easily from one place to another; often you cannot. To figure out whether a photograph is supplemented by a video clip (grainy, tiny and decidedly unspectacular), you must move the cursor over the photo and see if the arrow changes to a camera. To reveal additional information, you may have to click on an icon of a camera, a clipboard or a collection of maps. A "Roadmap" chart and a "backtrack" button offer modest navigational assistance.

The publisher says the disk contains about 1,400 photos, the book about 360. The CD-ROM also includes other information missing from the book, like responses to the questionnaires that the

I sit, mouse gripped firmly in one hand, and contemplate the situation. Shall I test the strength of our relationship and try to use the "penis tool" now? Or play it safe a little longer with the "hand tool"? Hmm. I click on the hand icon, then on a part of her torso.

Two male arms move into the video frame on my computer screen. Sierra moans as the hands touch her flesh, and she massages them into her breasts. They're not my hands. They're some other guy's hands. A pornographic CD-ROM male forearm-model is fondling Sierra's breasts for me.

113

GO

I'm with you in spirit, accompanied by the usual feelings of moral inadequacy and addictive faxophone.

Nicholson Baker
Berkeley, California

families answered, slide shows narrated by Charles Kuralt, overview maps of each country and the opening bars of national anthems.

But the book can display far more information at once, and uses that advantage to create a much stronger editorial viewpoint. In the book, each "big picture" is displayed alongside a caption describing the family and its possessions. The CD-ROM uses tiny "pages" of short captions that identify only three or four items at a time and even then obscure part of the picture. However, the CD-ROM's lists do include possessions not included in the photos; in the book, these are inconveniently noted on a page at the back.

What takes a couple of glances in the book can require many mouse clicks with the CD. The book's double-page "material world at a glance" table brings together a wealth of statistical and factual information. The disk omits some of this data entirely and takes 36 separate screens to display the rest. Although the CD-ROM can show the data as bar graphs, only one-third of the countries can be displayed at once, and then only alphabetically so statistics for Albania and Vietnam cannot appear simultaneously. Enlightening and amusing two-page spreads of 16 small photos of meals, televisions and toilets around the world appear as interludes in the book. Their equivalent in the CD-ROM includes additional photos as well as sections on toys, music, recreation, animals, transportation, homes, markets, kitchens, schools and most valuable possessions.

Unfortunately, the small format, low resolution and lack of captions make these photos tantalizing rather than useful.

The book lacks the simplistic introduction narrated by Mr. Kuralt. The disk lacks the book's detailed source notes and three introductory articles, including a comment on methodology in which Mr. Menzel cheerfully admits that the selection process included countries "I wanted to see" and an essay in which the historian Paul Kennedy points out the importance of understanding the project's findings "especially on a *comparative* basis," an aspect in which the CD-ROM comes up particularly short.

THE CD-ROM costs about $40, the book about $30. The book does not come with a number you can call for technical help.

DILBERT by Scott Adams

As It Is

DORIANNE LAUX

The man I love hates technology, hates
that he's forced to use it: telephones
and microfilm, air conditioning,
car radios and the occasional fax.
He wishes he lived in the old world,
sitting on a stump carving a clothespin
or a spoon. He wants to go back, slip
like lint into his great-great grandfather's
pocket, reborn as a pilgrim, a peasant,
a dirt farmer hoeing his uneven rows.
He walks when he can, through the hills
behind his house, his dogs panting beside him
like small steam engines. He's delighted
by the sun's slow and simple
descent, the complicated machinery
of his own body. I would have loved him
in any era, in any dark age; I would take him
into the twilight and unwind him, slide
my fingers through his hair and pull him
to his knees. As it is, this afternoon, late
in the twentieth century, I sit on a chair
in the kitchen with my keys in my lap, pressing

the black button on the answering machine
over and over, listening to his message,
his voice strung along the wires outside my window
where the birds balance themselves
and stare off into the trees, thinking
even in the farthest future, in the most
distant universe, I would have recognized
this voice, refracted, as it would be, like light
from some small, uncharted star.

— ✐ —

Attend any conference on telecommunications or computer technology, and you will be attending a celebration of innovative machinery that generates, stores, and distributes more information, more conveniently, at greater speeds than ever before. To the question "What problem does the information solve?" the answer is usually "How to generate, store, and distribute more information, more conveniently, at greater speeds than ever before."

    NEIL POSTMAN

The [digital] present is more frightening than any imaginable future I might dream up. If Marshall McLuhan were alive today, he'd have a nervous breakdown.

    WILLIAM GIBSON, AUTHOR
    NEUROMANCER, 1984
    FIRST TO USE THE TERM "CYBERSPACE"

MEDICAL EMERGENCY

TESTIMONY BY
PETER H. GOTT, M.D.

I'M DESPERATELY fighting the electronic revolution. I still use rotary telephones, disdain a cellular phone (because I welcome the peace and quiet in my automobile) and prefer to write my columns longhand (instead of having to pay attention to spelling and grammatical errors on a word-processor).

But, I fear, my one-man guerrilla warfare is doomed to failure: Electronic gadgets are everywhere. They're—aaaargh!—universally accepted, and—ugh!—becoming an integral part of modern medical practice.

Why, just the other day I called a colleague after hours to discuss a patient with a serious problem.

I ended up with a serious problem.

ON THE FOURTH ring, my call was answered automatically and I was told by a bland, unfamiliar, computerized voice that the office was closed. I knew this but the voice went on to describe a whole series of unwelcome options that I could enjoy by punching the buttons on my non-existent Touch-Tone phone.

If I had a question about a bill, I should push "1." If I wanted an appointment, I should push "2." If I wished to relay insurance information, I should push "3." And so on, until, if I needed to speak to the doctor, I should push "0."

By now, I was intrigued because I knew that this specialist—who had trimmed unnecessary financial fat from his overhead—had one secretary bookkeeper to help him in his office. This was a two-person operation. Why did it need 10 different telephone options when one—the direct line to the secretary—would take care of all my needs?

I waited—because I didn't have an "O" to push.

EVENTUALLY, there was a far-away whirring and a person—a real, live person—came on the line. He listened to my request and responded by saying he would "beep" the physician. I hung up. A few minutes later, the specialist called me back and answered my questions. Transaction complete.

But why, I wondered, did the doctor require such a production? All I wanted was a small audition; instead, I got an operetta in full dress with a philharmonic orchestra. Does this new push-button miracle make life easier for the consumer or is it really another ploy to separate doctors from their patients?

Then, of course, I speculated about how this system will probably germinate into a hardy communications weed that will enlarge uncontrollably and eventually besmirch the smooth putting green of medical practice.

PRETEND for a moment it's the year 2000. You pick up the (Touch-Tone, cordless) telephone to ask your doctor why he hasn't reported to you about a cholesterol blood test you had two weeks ago. On the fourth ring, a machine cuts in and gives you these directions.

Dr. Mergatroyd cannot come to the phone right now.

If you have an upset stomach, push 1.

If you have a cold, push 2.

If you have chest pain, push 3 and the pound key.

If you have a kidney problem, push 4.

If you are depressed, push 5.

If you have a sexually transmitted disease, push 6.

If you have labor pains, push 7.

If you have a fever and feel awful, push 8.

If you need brain surgery, push 9.

If you have a broken bone, push O.

If you want to beep the doctor, pay a bill, make a golf date or offer a dinner invitation, push "star," followed by the pound key.

If you have a rotary phone, too bad. Push off.

Well, it'll probably never happen this way. But who says it can't?

Scribble, Scribble, Eh, Mr. Toad?

LANCE MORROW

NATIONAL HANDWRITING DAY deserves to be celebrated with an updating of a part of *The Wind in the Willows*, a new chapter in the life of Toad of Toad Hall:

TOAD GAVE UP pen and pencil years ago, when he discovered the Smith-Corona manual portable typewriter. Toad loved his Smith-Corona. He played upon it like a flamboyant pianist. Now he massaged the keyboard tenderly through a quiet phrase, now he banged it operatically, thundering along to the chinging bell at the end of the line, where his left arm would abruptly fire into midair with a flourish and fling home the carriage return.

If Toad ever put pen to paper, it was reluctantly, to scribble in the margin of a college textbook ("Hmmmmm" or "Sez who?" or "Ha!"), or to write a check. Over the years, Toad's handwriting atrophied, until it was almost illegible. Who cared? Sonatas of language, symphonies, flowed from the Smith-Corona.

At length, Toad moved on to an electric model, an IBM Selectric, and grew more rapturous still. Toad said the machine was like a small private printing press: the thoughts shot from his brain through his fingers and directly into flawless print.

Then one winter afternoon, Toad came upon the marvel that changed his life forever. Toad found the word processor. It was to his Selectric as a Ferrari to a gypsy's cart. Toad now thought that his old writing machines were clattering relics of the Industrial Revolution.

Toad processed words like a demon. His fingers flew across the keys, and the words arrayed themselves on a magic screen before him. Here was a miracle that imitated the very motions of his brain, that teleported paragraphs here and there—no, *there!*—as quickly as a mind flicking through alternatives. Prose with the speed of light, and lighter than air! Toad could lift 10 lbs. of verbiage, at a whim, from his first page and transport it to the last, and then (hmmm), back again.

A happy life, until one day, Toad, when riding his bicycle in the park, took a disastrous spill. Left thumb broken, arm turned to fossil in a cast, out of which his fingers twiddled uselessly, Toad faced the future. He tried one-handing his word processor, his hand jerking over the keyboard like a chicken in a barnyard.

It was no use. There is no going back in pleasure. "Bother!" said Toad. He picked up a No. 1 Eberhard Faber pencil. He eyed it with the despair of a suddenly toothless gourmand confronting a life of strained carrots and peas. He found a schoolboy's lined notebook and started to write.

The words came haltingly, in misshapen clusters. Toad's fingers lunged and jabbed and oversteered. When he paused to reread a sentence, he found that he could not decipher it. The language came out Etruscan.

Yet Toad perforce persisted. It had been years since he had formally and respectfully addressed blank paper with only pen or pencil in hand. He felt unarmed, vulnerable. He thought of final exams long years ago—the fields of rustling blue-book pages, the universal low, frantic scratching of pens, the smell of sour collegiate anguish.

Toad drove his pencil onward. Grudgingly, he thought, This is rather interesting. His handwriting, spasmodic at first, began to settle after a time into rhythmic, regular strokes, growing stronger, like an oarsman on a long haul.

Words come differently this way, thought Toad. To write a word is to make a thought an object. A thought flying around like electrons in the atmosphere of the

Is it OK if I use a mechanical pencil?

Fred Bonavita
San Antonio, Texas

I use a computer. This enables me to be highly efficient. Suppose, for example, that I need to fill up column space by writing BOOGER BOOGER BOOGER BOOGER BOOGER. *To accomplish this in the old precomputer days, I would have had to type "BOOGER" five times manually. But now all I have to do is type it once, then simply hold the left-hand "mouse" button down while "dragging" the "mouse" so that the "cursor" moves over the text that I wish to "select;" then release the left-hand "mouse" button and position the "cursor" over the "Edit" heading on the "menu bar;" then click*

123

brain suddenly coalesces into an object on the page (or computer screen). But when written longhand, the word is a differently and more personally style object than when it is arrayed in linear file, each R like every other R. It is not an art form, God knows, in Toad script, not Japanese calligraphy. Printed (typed) words march in uniform, standardized, cloned shapes done by assembly line. But now, thought Toad, as I write this down in pencil, the words look like ragtag militia, irregulars shambling across the page, out of step, sloven but distinctive.

Toad reflected. What he saw on the penciled page was himself, all right, not just the content of the words but the physical shape and form of thought. Some writers do not like to see so much of themselves on the paper and prefer to objectify the word through a writing machine. Toad for a moment accused himself of sentimentalizing handwriting, as if it were home-baked bread or hand-cranked ice cream. He accused himself of erecting a cathedral of enthusiasm around his handicap.

At length Toad could see his own changes of mood in the handwriting. He could read haste when he had hurried. He thought that handwriting would make a fine lie-detector test, or foolproof drunkometer. Handwriting is civilization's casual encephalogram.

Writing in longhand does change one's style, Toad came to believe, subtle change, of pace, of rhythm. Sentences in longhand seemed to take on some of the sinuosities of script. As he read his pages, Toad considered: The whole toad is captured here. *L'écriture, c'est l'homme* (Handwriting is the man). Or: *L'écriture c'est le crapaud* (Handwriting is the toad). What collectors pay for is the great writer's manuscript, the relic of his actual touch, like a saint's bone or lock of hair. What will we pay in future years for a great writer's computer printouts? All the evidence of his emendations, his confusions and moods, will have vanished into hyperspace, shot there by the Delete key.

Toad found himself seduced, in love, scribbling away in the transports of a new passion. Toad was always a fanatic, of course, an absolutist. He bought the fanciest fountain pen. His word processor went first into a corner, then into a closet with the old IBM.

Toad thought of Henry James. For

the left-hand "mouse" button to reveal the "edit menu;" then position the "cursor" over the "Copy" command; then click the left-hand "mouse" button; then move the "cursor" to the point where I wish to insert the "selected" text, then click the left-hand "mouse" button; then position the "cursor" over the "Edit" heading on the "menu bar" again; then click the left-hand "mouse" button to reveal the "edit menu;" then position the "cursor" over the "Paste" command; then click the left-hand "mouse" button four times; and then, as the French say, "voila!" (Literally, "My hand hurts!")

If you need this kind of efficiency in your life, you should get a computer.

DAVE BARRY

125

decades, James wandered Europe and the U.S., staying in hotels or in friends' houses. He was completely mobile. He needed only pen and paper to write his usual six hours a day. Then in middle age, he got writer's cramp. He bought a typewriter, and, of course, needed a servant to operate the thing. So now James was more and more confined to his home in Sussex, pacing the room, dictating to the typist and the clacking machine. James became a prisoner of progress.

Toad, liberated, bounded off in the other direction. Light of heart, he took to the open road, encumbered by nothing heavier than a notebook and a pen. Pausing on a hilltop now and then he wrote long letters to Ratty and Mole, and folded them into the shape of paper airplanes, and sent them sailing off on the breeze.

———— ✐ ————

What I find disturbing is an insidious tendency in my profession to put computerization and high tech information systems above the values of good professional intervention. What follows is a hollow, cost-efficient methodology for processing "caseload." There is no high tech device which will ever replace the bond of verbal communication face-to-face.

Glena Dean, social worker
Livermore, California

THE COMPULSION TO THINK

TESTIMONY BY
ALAN DAVID SOPHRIN

KNOWING THAT ONE DAY I will be punished for my mortal sin, I suffer more from anger than from fear, not because my punishment will not be just, (it will) but because it will not be swift. When my day of judgment comes, I will be consigned to a long and painful agony in technology purgatory.

My sin, which the gods of technology cannot forgive, is that I have chosen to live in a place where there are so few resident humans that it will not be profitable to include me in the technology of the information age. I know this is so because now it is not even profitable to include me among the recipients of cable TV. If there is no profit in hooking me to cable TV, certainly there will be no profit in hooking me to the devices of the information age when it arrives. When it does, everyone in the world (except me and a few like me) will be a paying customer in a vast network of fiber optics and advanced switching that will bring them all the information there is about everything.

At present, I am not totally uninformed. I receive newspapers and noncable free network TV, but in those few moments when I face reality, I know my days with these things are numbered. As we have more and more mergers of the huge enterprises that control all the wire and wireless transmissions of information, the day will come when one or two of these enterprises will control all communications everywhere. Then there will be no more free TV and no more newspapers. Also, there will be no more books because assimilating all the vast amounts of information the communications giant (or giants) will sell to us (except to me) will not leave anyone time to read.

I concede that I have enough books on hand to occupy my mind after I go to technology purgatory, but I know the gods of technology do not intend to punish me with boredom. (Present TV programs, cable and non-cable, have established that boredom is no longer punishment.) Instead, my punishment will be that I will not be relieved of the human compulsion to think. Thinking has always been painful so the primary benefit to those who will be given the privilege of buying the new age's great mass of information will be that they will not have either the time for or the need for thought. In this new age, information will replace thought. So my punishment, as a non-information customer, will be to live out my days suffering the pain of thinking.

I AM NOT asking for sympathy. I know that all I would have to do to stay out of technology purgatory would be to move to a more populated place. But I am too weak in character to throw off my addiction to the forests and lakes and hills where I live. So I stay and await my fate.

My Son Saves Nanoseconds

TESTIMONY BY
ARTHUR HOPPE

I WAS LEAFING through *Newsweek*'s cover story, "Exhausted!" while having breakfast in the kitchen. By golly, says this worried magazine, we Americans are "fried by work (and) frazzled by lack of time."

Darned if I could figure out why the younger generation should lead such frenetic lives. But just then my son, Mordred, popped in the door with an application for a mortgage for me to co-sign.

"I don't see how you live without a fax machine, Dad," he said, impatiently drumming his fingers on the kitchen table as I hunted for a pen. "I could've saved 20 minutes not having to drive over here."

"Always glad to see you, Mordred," I said.

"You and Mom don't even have voice mail," he complained as he looked over my shoulder. "If I want to leave you a message, I have to call three times to catch you home, and then you invariably want to chat. There goes another ten minutes out of my day."

"Always nice to talk to you, Mordred," I said. "I guess your mother and I aren't much for these modern time-saving devices."

"You've got to get with it, Dad," he said. "Take a little thing like the

E-mail we've got down at the office. I save a good half hour a day not having to constantly wander over to somebody else's desk to ask a question. Give it the old rat-a-tat-tat on my computer and that's that."

"Will wonders never . . ." I was interrupted by a raucous "Beep! Beep! Beep!"

"Whoops, that's my beeper," said Mordred, pulling the device out of a holder on his belt. "Saves me having to check in with my office every ten minutes."

"Do you talk to it?" I asked.

Mordred laughed. "See, it's got the number calling me on its digital screen. I just pick up a phone and call whoever it is back."

"What if you were lying on a beach somewhere sipping a mai tai?" I asked. "You'd have to spend half an hour looking for a phone."

"Not if you've got a cellular," he said, triumphantly extracting the latest model from his hip pocket. "Greatest time-saver since call waiting and call forwarding. Say the big shots are looking for me for a conference call . . ."

He rapidly poked the requisite buttons and listened. "No, I'm not interested in 50 percent off 3-by-5 photos of my family today or any other day," he said testily, clapping the phone together and stuffing it away.

"I don't think I'm technologically capable of operating a time-saver like that," I said.

He frowned. "Look, Dad, I don't mean to harp, but the least you and Mom could do is put in a modem and a computer in the cabin up at Lake Algae. Think of the time I could save messaging you on the Internet instead of having to sit down and write you a letter whenever I wanted to know when the place was free."

"It's the only time we ever get a real letter from you, Mordred," I said. "But why don't you sit down and have a grapefruit?"

"A grapefruit?" he asked incredulously. "Do you realize it takes a good three minutes to eat a grapefruit? I haven't had a grapefruit in years."

So I gave him a banana and sent him on his way, hoping that my son would not prove to be yet another flameout of the new time-saving generation.

One upsmanship at our house is telling someone else how many more new gadgets you don't have than anyone else. . . . My motto: A pencil sharpener in every room.

Elinore Leva
Bodega Bay,
California

The oversell on the "information superhighway" exploits the same public gullibility that true atomic-energy believers exploited decades ago. It's a gullibility that flows from a touchingly credulous eagerness to believe that new miracle ages are constantly lurking just around the corner.

RUSSELL BAKER

I'm a shunpiker myself— Victorian word for "traveling the byways and avoiding the turnpikes."

Harriet
Eugene, Oregon

A Read-Only Man In An Interactive Age

TESTIMONY BY
GEORGE FELTON

AMERICA'S OBSESSION with interactivity is going too far. Recently my CBS affiliate, WBNS-TV, wanted me to tell them what the news would be: "The future of television news is in your hands. You pick the story you want to see tonight at 11." I was given three options, each with an 800 number: 1. foods that make you nuts, 2. cabbies—is the fare fair? 3. things to know before you sign a house contract. I was supposed to jump to the phone and vote.

This is just a gimmick, of course. It's condescending, over-hyped, and irrelevant all at once, but none of that bothers me; local news as a genre is virtually defined by such terms. What does bother me is the growing demand that I stick my finger into the middle of every experience I have, no matter with whom or what. In this instance, I want WBNS to pick the news. I've been doing my job all day. Let them do theirs.

Everywhere I turn I'm being urged to dial, punch, vote, fax, E-mail, double click, or simply stand up, grab the microphone, and let loose with my comment, my question, my need, my self. It's our new, interactive way of being, and if it's all about control, I don't think I want that much.

Interactivity's key premise is that, at long last, I get to direct the action: I point and click or stand up and fire my salvo, after which the thing viewed changes its course for me. I'm told I'll soon be able to sit in my living room and press a button routing the movie/book/video/CD/ experience-mechanism in the direction I want it to go. Why let Thoreau lead me around by the nose if I can hypertext my own way across Walden? Why move my lips if I can press a button and have James Earl Jones move them for me? Why suffer Walden's dreary, gray type at all if I can punch up on CD-ROM a video of the more recent, more dramatic fight to preserve the pond from developers, or better yet, watch a Don Henley benefit concert that makes me feel noble about what I just quit reading?

Which gets me to one of my problems with the interactive future: When I'm finally free to direct where *everything* goes, I'll never go anywhere I don't intend. In fact, I'll never learn anything new, just keep recycling a few of my favorite things. I do enough of this already. I don't need invitations to spiral even deeper into my own black holes.

Nor do I need to have a "conversation" with Thoreau in which I determine what's interesting and get appropriate text bytes in response. If it took him two years to live the book, nine years to write it, and six drafts to get it right, I can at least shut up and let him determine what's interesting But this has become an old-fashioned idea, very out of tune with the noisy, nosy I-Me-My-isms of interactive life.

My Lutheran church, for example, no longer finds the call and refrain of its own liturgy sufficiently interactive: Now the pastor opens with a warm-up monologue during which he invites comment. We interrupt the service to stand, shake hands, and exchange greetings. The kids are trooped up front for an interactive mini-sermon—the pastor telling them a parable and asking questions, during which they fidget and mumble. I wonder if next week they'll be asked to field questions from us.

I go to an art exhibit, only to discover it's nothing without me. I must help create the art by intersecting it, walking into the viewerscope of a camera, spinning a globe and making videos reel, sticking my nose in a little hole and having a woman in there talk back to me. As one art critic said about the show, it's hard to know what to make of it, but it sure is noisy.

At home my newspaper thinks if I'm just reading it, I'm not doing enough. I've got to say how the mayor's doing, call the tax hot line with my Schedule C problems, fill in today's clip-out quiz about O.J. I turn on the radio, but it turns on me. Ron Barr's "Sports Byline USA" has a line open and needs a call. Our local NPR station is schmoozing with P.J. O'Rourke and wants me to make it a three-way. I click on TV—and not a moment too soon. VH-1 says it's now or never if I want good Tom Petty tickets for the summer tour. Bob Berkowitz needs some fast answers about masturbation. NBC's "Dateline" doesn't know what to make of the latest news scandal. Can I take a quick gut check and give them a call, send a fax, maybe an E-mail?

I reach for a beer, but even it wants to hear from me. It says so on the little paper band around its neck, the band I love to rake off with my thumbnail while drinking, the one that used to say "the champagne of bottled beer" or some other satisfyingly dumb advertising slogan. But now I look down and see an 800 number. I am being urged to speak with my beer, which is entirely different from mumbling in its presence, a far more traditional interaction and, these days, perhaps a more meaningful one

It is not just crankish and extreme to say that a "kind of massacre" is going on in libraries right now. There is the exuberant recycling of the card catalogues themselves; and then there is the additional random loss of thousands of books as a result of clerical errors committed in disassembling each card catalogue, sorting and boxing and labeling its cards, and converting them en masse to machine-readable form—a kind of incidental book burning that is without flames or crowds and strangest of all, without motive.

135

NICHOLSON BAKER
THE NEW YORKER

THE FETISH OF IMPERMANENCE

D.T. MAX

IN THE HEART of official Washington, D.C., down the street from the Capitol and at the same intersection as the Supreme Court and the Library of Congress, stands an incongruous statue of Puck, whom the *Oxford Companion to English Literature,* soon to be issued on CD-ROM, defines as "a goblin," and whom Microsoft Encarta passes over in favor of "puck," which it defines solely as a mouselike device with crosshairs printed on it, used in engineering applications. The 1930s building next to the statue is the Folger Shakespeare Library. Two flights below the reading room, designed in the style of a Tudor banquet hall, next to which librarians and scholars click quietly on laptops and log on to the Internet's Shaksper reference group for the latest scholarly chatter, is a locked bank gate. Behind it is what librarians call a "short-title catalogue vault"—in other words, a very-rare-book room. This main room—there is another—is rectangular, carpeted in red, and kept permanently at 68 degrees. Sprinkler valves are interspersed among eight evenly spaced shelves of books dating from 1475 to 1640 and lit by harsh institutional light. Of these books 180 are the only copies of their titles left in the world: you can spot them by the small blue slips reading "Unique" which modestly poke out from their tops. At the end of the room is a long shelf on which stacks of oversize volumes rest on their sides: these are nearly a third of the surviving First Folio editions of the plays of William Shakespeare. When the First Folios were printed, in the 1620s, printing was still an inexact art. Each page had to be checked by hand, and the volumes are full of mistakes: backward type, ill-cut pages, and variant lines. Several copies lack the 1602 tragedy *Troilus and Cressida*, owing to a copyright dispute. And yet, 370 years after they came off the printing press, you can still pull down these books and read them. The pages are often lightly cockled and foxed, because the folio was printed on mid-priced rag paper, but the type is still bright and the volume falls open easily. You can balance it on your lap and run your finger along the page to feel the paper grain in that sensuous gesture known to centuries of book readers: here is knowledge.

In 1620 Francis Bacon ranked printing, along with gunpowder and the compass, as one of the three inventions that had "changed the appearance and state of the whole world." Indeed, the existence of multiple identical copies of texts that are nearly indelibly recorded, permanently retrievable, and widely decipherable has determined so much of modern history that what the world would be like without printing can only be guessed at. More books likely came into existence in the fifty years after the Gutenberg Bible than in the millennium that preceded it. "Printing was a huge change for Western culture," says Paul Saffo, who studies the effect of technology on society at the Institute for the Future, in Menlo Park (where the receptionist also uses an IBM Selectric).

"The dominant intellectual skill before the age of print was the art of memory." And now we may be going back.

For the question may not be whether, given enough time, CD-ROMs and the Internet can replace books, but whether they should. Ours is a culture that has made a fetish of impermanence. Paperbacks disintegrate, Polaroids fade, video images wear out. Perhaps the first novel ever written specifically to be read on a computer and to take advantage of the concept of hypertext—the structuring of written passages to allow the reader to take different paths through the story— was Rob Swigart's *Portal,* published in 1986 and designed for the Apple Macintosh, among other computers of its day. The Apple Macintosh was superseded months later by the more sophisticated Macintosh SE, which, according to Swigart, could not run his hypertext novel. Over time people threw out their old computers (fewer and fewer new programs could be run on them), and so *Portal* became for the most part unreadable. A similar fate will befall literary works of the future if they are committed not to paper but to transitional technology like diskettes, CD-ROMs, and Unix tapes—candidates, with eight-track tapes, Betamax, and the Apple Macintosh, for rapid obscurity. "It's not clear, with fifty incompatible standards around, what will survive," says Ted Nelson, the computer pioneer, who has grown disenchanted with the forces commercializing the Internet. "The so-called information age is really the age of information lost." Software companies don't care—early moviemakers didn't worry that they were filming on volatile stock. In a graphic dramatization of this mad dash to obsolescence, in 1992 the author William Gibson, who coined the term "cyberspace," created an autobiographical story on computer disc called "Agrippa." "Agrippa" is encoded to erase itself entirely as the purchaser plays the story. Only thirty-five copies were printed, and those who bought it left it intact. One copy was somehow pirated and sent out onto the Internet, where anyone could copy it. Many users did, but who and where is not consistently indexed, nor are the copies permanent—the Internet is anarchic. "The original disc is already almost obsolete on Macintoshes," says Kevin Begos, the publisher of "Agrippa." "Within four or five years it will get very hard to find a machine that will run it." Collectors will soon find Gibson's story gone before they can destroy it themselves.

A few years ago on a walk across England, my local companion pointed to a hill in the Borrowdale Valley in the Lake District and explained that 150 years ago (give or take) a storm there had caused a landslide which exposed graphite for the first time. It was considered so valuable that it was kept under lock and key by an armed guard. My lodging in the nearby town of Keswick was actually a converted pencil factory.

Alice C. Corley
Santa Rosa, California

Personal computers are notorious for having a half-life of about two years. In scientific terms, this means that two years after you buy the computer, half of your friends will sneer at you for having an outdated machine.

PETER H. LEWIS

139

DON'T FAX ME IN

CAROL ANN MESSECAR

The time has come to grind this axe,
I'm sick and tired of good old FAX.

Every drone on the phone tries to prove that he's better,
By faxing me a perfectly ordinary letter.

So I slog on over to my neighborhood copier,
And there's my fuzzy letter, it couldn't be sloppier.

It's like real estate, they say, "You mean you don't OWN?"
"Why, I've one fax in my office and one in my home."

Well, I caved in and joined the fax extravaganza,
To rush-rush the most humdrum of memoranda.

People in L.A. get themselves in a spinner,
If New Yorkers don't jump up and read that fax with dinner.

It's a real time-saver to spend an hour by the phone,
Listening to someone else's fax busy-tone.

Now any office snoop can read each other's contracts and pay,
And the detail of deals they've made that day.

"Warning" says a lawyer's faxy cover letter speed-it,
"If this fax isn't for you, then please don't read it."

Maybe I'm the only one left in town,
Who thinks "luxury" means "the time to slow down."

And "privacy," that's a word that I fear
Our technological toys will make disappear.

Because after FAX and E-mail—what? It sure could spoil it.
Someone might invent a copier to attach to the toilet.

Or a video-phone where you can't help but see,
Rumpled folks in the shower or in bed, maybe.

Oh, give me the mailman whom I've come to trust,
But don't fax me in.
Give me Fed-Ex if you really must,
But don't fax me in.

141

Let me wander while I ponder a world-without-beeps.
For I've miles to go before I sleep.

And a well-thunk thought takes time to get right.
So don't fax it to me. Sleep on it tonight

CUI BONO?

EVA T. H. BRANN

THIS LITTLE MEMORANDUM from the foot-dragging rear-guard of the electronic revolution is in praise of writing on paper and reading from books. It rests lightly on two large assumptions.

The first answers to the Lead Pencil Club watchword "Not so fast." It says that the flux of human events is in some respects cyclical, so that the hindmost, far from being taken by the devil of irrelevance, find themselves now and then out front with our better angels of repentance. Put as an encouragement: Stick in the mud and the returning tide of good sense will float you off first.

The second assumption heeds the cry "Not so much" (found on the pediment of the Temple of Apollo at Delphi in the form *Meden agan*, "nothing in excess"). It claims that the returning vortices are borne along in a stream whose law of progression—not to be confused with Progress—is quantitative increase. Some human institutions (they scarcely need instancing) show a near-irresistible tendency to get bigger without an accompanying propensity for getting better. Hence cavillers and doubters, far from being "out of it," are very much "with it," providing "it" is properly defined, namely as the substance of life. Put as an entitlement: You have a right to query the praise of increase and multiplication by means of that old conversation-stopper of a question: *Cui bono?* "To what good purpose?"

They say that we live in a world of vastly accelerated change, sweeping innovation, and unprecedented revolution, which will bring all the world and its goods instantly—though for the most part "virtually"—to us. Well, perhaps—though those of us who recall the last bout of millennial futurology, the one in the seventies, when the universal prosperity of the peaceful global village was going to make the psychological management of leisure the great problem of the age, may wonder. Let us suppose, nonetheless, that some approximation to the prophesied jubilee of obsolescence will in fact occur, so that we shall live in a world of paperless offices, bookless libraries, limitless information, lagless communication, wall-less schools, and teacherless classrooms.

Here is my question: Will anything have happened? I mean, will there have been—size, speed, and displacement aside—some access of human good, some increase of life's substance?

It is a fair, not a merely curmudgeonly question. For the promised millennium is a *media* millennium. "Medium," recall, signifies "means," and means are nothing without their ends.

But no, not nothing. For end-less means take on a curious life of their own; lacking the derivative dignity that is proper to them they establish themselves, rebelliously, in their own right. A serious argument can be made—though not here—that such transmogrifications of the intrinsically secondary into the derivative primary have been characteristic of modernity from its beginning and are escalated in its post-modern sequel: Relations go into terms, image is taken for original, quantity turns into quality, second nature supersedes grown nature, and so also the medium becomes the message. Or, seen from the opposite perspective: We live with a tendency away from what is first, intrinsic, and absolute, toward what is derivative, extraneous, and relative—from service to administration, workshop to bureau, salvation to opportunity, and so also from significance to dissemination.

This great litany of remotion from the given world poses an old theological question in a new secular form. The old issue was whether the devil's deviltry was a privation of good or a positive evil, a something or a nothing. The new question is whether these electronic benefactions are positive goods intrinsically conducive to our happiness or nothing in themselves beyond an opportunity. It happens to be a question not often or loudly asked. The paeans of praise are pretty universal.

Nonetheless, I mean in this note not to deal with profound issues, but with marginal and superficial matters, matters of the sensibility. These are the light, subtle, peripheral losses the electronic revolution will bring. They are aesthetic defaults, "aesthetic" taken in its original meaning of "sensory," as in sensory deprivation. These pleasures and refreshments of the senses are evanescent and sometimes below the threshold of a busy awareness, but they sum up into a way of being human.

From Writing to Processing

HERE IS A LOVELY IMAGE: The banks of a limpid pool fringed with papyrus plants, and a scribe sitting on its banks painting on a book roll words of prospectively ancient wisdom. I am not, however, proposing a return to papyrus, the historical and etymological ancestor of our paper, or to a brush in favor of our lead pencil. For I think that anything is possible, that deep wisdom can be read off a papyrus being unrolled or an electronic text being scrolled and that a great text can be composed on a parchment or a word processor. Processing, which does so little for cheese, may do much for literature. It is possible, but there are misgivings.

Those who can't afford the new technology, or don't master its use, will become information-poor. Millions of people could become "illiterates." Indeed, if you consider illiteracy a serious problem today, just wait. This new technology revolution in information access and retrieval promises to give new meaning to the word.

FUND-RAISING LETTER
(FOR COMPUTERS)
NEW YORK PUBLIC
LIBRARY

145

I remember former fellow employees who were once human and could carry on an interesting conversation, ●●●●sionally read and discuss a good book, go on a picnic, etc., and are now what I refer to as "computer-outted!" They have become dull, stare into space; have a newly acquired zombie like existence, and are clueless about ✏

A usage note in my favorite dictionary (the *Heritage*), says that a processor is "an apparatus for preparing, treating, or converting material: *A wood pulp processor.*" The entry for word processing amply compensates for these intimations that what is done to wood might happen to fiction. It credits the process with "the *creation* (my italics), input, editing, and production of documents and texts by means of computer systems." Here we have the process as poet; writers, in turning on the machine, will be as latterday Pygmalions or Frankensteins, galvanizing the inanimate machine into a productive life to supplant theirs. But this is hyperbole. I am interested in small supersessions.

And first of these is, I think, the loss of the cursive motion, the running of the lead over paper, leaving behind a trace of its own substance. There is an incident satisfaction in coming to the end of the lead, with the exhausted sheaths remaining as a tally of an accomplishment more sober and more lasting than that betokened by those souvenir collections of wine bottle corks and concert ticket stubs.

Cursive continuity is opposed to digital discreteness in being analogous to mental speech, which runs smoothly in words and sentences rather than in separate phonemes. Words, even sentences, have continuous pitch and wandering beat, unlike the discrete tones and fixed measures of music, which are, except for hellish or mawkish glissandos, well represented by separate notes joined at most by their flags. Although early writing consisted of scratched capitals—the Greek word for letter, *grámma*, seems to be onomatopoetic—and often appears to express a sheer delight in lettering and in the discovery of these least L–M–N–ta of speech, our latterday, literate, non-monumental intimacy with visualized speech is best represented in the cursive hands. It is cursive as it comes from the hand that expresses individual character and mimics the mood of the day—scrawlly in afflatus, tight and tiny in drought—a private log of life's visicitudes.

Writing, too, has the advantage of limitless symbolic invention, of arcane private signs, abbreviations, and pictographs not on any keyboard. This boon is part of the immediacy of the writing motion, which is not capturable even by the electronic stylus, that sense of being literally on top of one's work, without any intervening processing black box.

what is happening anywhere except on the face of their screens. A really curious phenomenon!

Gladys Evan
Clearwater, Florida

Virtual reality points to a boundless capacity for deception. Not simply by governments or corporations, but by hostile individuals acting on each other. We can do this today, but we are increasing the sophistication of deception faster than the technology of verification.

147

The consequence of that is the end of truth. The dark side of the information technology explosion is that it will breed a population that believes nothing and, perhaps even more dangerous, a population ready to believe only one "truth" fanatically and willing to kill for.

ALVIN TOFFLER

Handwriting, manu-script proper, has, moreover, other physiological advantages over pecking away at keys, quite apart from the well-known strains on wrist and back. A paper tablet can, without fear of electrocution or malfunction, be taken into that imagination-releasing homeostatic venue, the bathtub, or into the earthly Eden-analogue, the garden (where even a lap top is a finicky intruder), or hither and thither, as the spirit lists. Feet can be up—who wants to be always chair-shaped while thinking?—and one hand can be free to fiddle with a strand of hair or to mark the needed pages in a book.

Then there is that wonderfully biddable medium, the paper tablet. It has the lovely property of *being there;* it is, as the philosophers say, a real substrate. A text on a screen is neither here nor there, since it is scrolled across an unfathomable void, untethered to material reality. It, the electronic text, moves, not the reader, in a curious reversal of natural motion and rest; it is a foretaste of a virtual world in which the human being is the minimally mobile monitor of a moving illusion.

The materiality of paper makes it the capable bearer of a text *together* with its genetic history. On paper there can be pentimentos and palimpsests, crossings and overwrites, erasures and marginalia, cuttings and pastings. Everything is there at once, stratified. The overlaid traces of rethinking have the interest of an archaeological dig, where the remains of an early foundation are visible within the final structure. Since much rewriting sometimes produces prose apparently written by a committee of one, it is good to have the record of an earlier spontaneity to return to.

But writing on paper produces the opposite effect as well. For the messiness of correction encourages that try at original perfection, that mental working over, that tasting on the tongue of words, before pencil is ever put to paper, which we know great stylists to have excelled in. Silent speech is, after all, as close to the mental conception as we can get, and it is in that internal workshop that words are most animate and sentences most melodious. Easy emendation obviates early perfection and the good impulse to crumple the whole mess up and start from scratch. Word processing poses the temptation to settle for transience and to tolerate patchwork. If the painted papyrus demanded first-time perfection, if the print-out allows indefinite

Something has got to be done! In the last three weeks I have thrown away two cordless phones, the TV remote stopped working, and I have to start unplugging the washing machine between every load because of another faulty wire. My three-year-old computer (at $3,500) is obsolete, and my VCR broke (at $210, the repairman told me to trash it—it is disposable!). At work I spend 35% of my time typing because I can't hire a secretary—they are obsolete. Hell, I'm 49 and I'm obsolete!

Kenn Young
Reno, Nevada

149

I am moreover a Luddite, in what I take to be the true and appropriate sense. I am not "against technology" so much as I am for community. When the choice is between the health of a community and ✏️➡

latitude, the pencilled paper provides a nice chance for first perfection and light later emendation.

In our day, of course, every manuscript does finally turn into a typescript. There is great value in that hiatus in which the work is out of our hands, being cast into formal and objective shape—an expectant and festive time, something like awaiting the reappearance of a friend who is changing into a tuxedo. And of course, the first glance shows that alterations are needed.

Finally, no inspiration committed to paper is lost by a sudden power outage. Paper is safe. (Though D. H. Lawrence *alleged* that he lost the first completed manuscript of *The Seven Pillars of Wisdom* by theft on a train, the story has been doubted.)

From Reading to Scanning

A COPSE OF BEECHES on a hillside is lovely, and so are beech leaves, from whose tough straight veins we used to make by defoliation such satisfying fish skeletons. The beech gave the book its name, from the beech branches on which runes were carved. It is fitting that books are said to have leaves; they are somehow akin to the woods which provide their matter and often their contents: "The woods are lovely, dark and deep . . ."

I do believe, once again, that a fine text is absolute and indefeasible, not to be done in by any mere format. A Bach concerto stays itself whether scored for harpsichord or piano, performed by an antique consort or an electronic synthesizer, interpreted metronomically or romantically. Perhaps the Bible will always seem to be most what its name betokens, *The* Book, when printed on gilt-edged onion skin, but for newspapers electronic transmission will be just great—no more newsprint on the Sunday doughnut and no more huge pages to wrestle down. It is ultimately all the same whether a text is read off graven tables or electronic tablets, if it be but read.

Well, maybe not *all* the same. Bound books do have physical and sensory distinctness. The preternatural molded plastic smoothness of the machine, on the other hand, goes with the Protean slipperiness of its display (except that the original Proteus of the *Odyssey* is a smelly old sea-beast, while a computer is as

technological innovation, I choose the health of the community. I would unhesitatingly destroy a machine before I would allow the machine to destroy my community.

WENDELL BERRY

I am a fifth generation American Indian (Delaware), author of two books and 25 published poems. These evolved first by pencil (on old envelopes), then pen, and finally drafts of this disgusting word processing machine.

Jeanne Rockwell
Barrytown, New York

The Beat Mind is already looking at the next Cold War, the war against computers. It's a war by humans against the nonhumans.

LAWRENCE
FERLINGHETTI

151

sanitized an object as can be). Books too, like paper tablets, can be carried anywhere and, not being delicate, will blend in anywhere. A good soaking may warp their covers but does not wipe out their message. In fact, loosened spines and frayed corners bear present witness to the preoccupations of past decades: *Jean Christophe* in the teens, the *Sonnets* in the twenties, the *Dialogues* and the *Critiques* in the thirties, forties and fifties, and now those *Odyssey*-derivatives, travel books.

You *can* tell a book by its cover—in fact that's how you do tell it. Books in their variety and singularity give each text, that "airy nothing, A local habitation and a name." Electronic texts are, of course, placeless; they all appear on the same screen, and when they're not there, they're nowhere.

In their solid separate physical presence "volumes," rightly so called, thus cause texts to be spatially external to each other. Hence they each have their place and position—assigned, expressive, and hierarchical. Some serviceable standbys like the *Thesaurus* live within reach of the ruling presence (oneself). Some unlovely tracts, retained only because it is contrary to nature to throw a book out, have to lie flat and in stacks on the uppermost shelves. Books for study live in the study, novels in the bedroom, pop-sociology in the living room. When work is in progress books lie open on the floor, the hassock, the chair arm, giving a panoramic view of the subject. A bonus is the exercise of running downstairs for a reference, climbing up a step-ladder; a home library is a veritable gymnasium—no being glued to a screen.

The ordering of books in a private library calls for real decision: Is the order to be handiness, affection, time of acquisition, subject? When I studied at Yale, there was a story abroad that a graduate student, employed to put the personal library of the defunct Roman historian Rostovtseff in order, arranged the volumes accurately by height, and in undecidable cases by color. It made a mad sort of sense to me.

Just as books all occupy their own voluminous place (which is of course what, in an age of publication explosion and indiscriminate acquisition, does them in), so each book separately has its own internal space. When opened to one place,

Gutenberg (hesitantly):
Perhaps the book, like God, is an idea some men will cling to. The revolution of print pursued a natural course. Like a river, print flowed to its readers, and the cheapness of the means permitted it, where the channel was narrow, to trickle. This electronic flood you describe has no banks; it massively delivers but what to whom? There is something intrinsically small about its content, compared to the genius of its workings. And—if I may point out a technical problem— its product never achieves autonomy from its means of delivery. A book can lie unread for a century, and all it needs to come to life is to be scanned by a literate brain. This CD-ROM of yours—what machine will be able to read it a hundred years from now? Each

153

all the others are still right there simultaneously stacked up in space, not strung out in time but here and now ready to hand.

Modernist fiction is characterized by critics as "spatial," meaning that its narrative is not temporally linear but perturbs times and locations, so that it hasn't been read until it is reread. But some such effect is to be found in most great fictions; they all demand back reference and anticipatory memory, a sort of simultaneous or parallel reading—in short, several fingers in the book. Think of the *Alexandria* or the *Raj Quartets*. To be sure, all this simultaneity might be obtainable on a screen, but it would be too huge for home use.

Books, furthermore, make wonderful files for the souvenirs of life. Clippings, poems, postcards resurface after years: "We went to Winchester Cathedral to see Jane Austen's tomb and thought of you." "Here is a sculpture of Pythagoras. Do you think it is a good likeness?" Where else would one keep all that stuff and be sure to come on it again someday? The same holds for the marginalia, underlinings, cross references. A well-read book comes to incorporate its own pencilled commentary.

All in all the scanning that a screen promotes does not seem quite the same as the reading a book invites—for dozens of evanescently small sensory causes, though in time some people will learn to "riddle" texts (for that is what to "read" means from way back) on a screen as attentively as they once did in books.

Yet it will always be easier to switch off an irritatingly formidable electronic text than it is to avoid a printed tome—a text not called up is but a title entry, but an unread book is a rectangular reproach.

And in general there is a real danger that in the conversion from book to disk libraries will bring about a benignly meant fireless book burning, the kind in which public libraries are now engaging as they weed out nice old novels to make room for multiple copies of the latest best seller. For what will be available will be the latest version, say of reference works, which is sometimes the worst. Take for example the *Thesaurus*, that treasure not of denotations but of connotations, where one could go to find the term for an aphasic thought or the word that had slipped the mind. The old Roget was organized on a wonderful yin and yang principle by synonyms and

generation of these machines destroys the previous; the very speed and momentum of your revolution erode its contact with the earth. You speak of this global Internet as if it transcended human brains; but man is still the measure of all things.

Gates *(collapsing with a hiss):* That can be fixed, eventually.

JOHN UPDIKE

155

This is what happens when you send electronic mail, as millions of Americans now do. Your fingers fly; so, in a sense, do your words; and you never run out of stamps. But if you're the recipient and the correspondence is personal, you are not getting all the message. ✏️ antonyms, so that you could snaffle a word through thinking of its opposite. Right in the center, no. 500, was "Sage" followed by "Fool" and around these was provided a universe for your orientation, running reverently but confidently from no. 1, "Existence" to no. 1000, "Temple." The revised system is arranged alphabetically, like a meaningless and sketchy dictionary, under the leading entry, and if that's the word you're missing, you're out of luck. Progress!

I nearly forgot to mention in favor of books that they make the liveliest wall coverings and very effective sound proofing.

From Experience To Information

THEY SAY THAT this is the Information Age and that this Electronic Revolution is an Information Revolution. They are evidently right.

"Information" here is not used in the older senses, as in "information of the mind," the imbuing of the intellect with noumenal shapes, or as in "a well-informed person," someone who is full of pertinent factual knowledge.

This information is rather an object to be stored and retrieved, subject to inputs, outputs, and updates. (The Germanic propensity for prepositional prefixes that computer language affects is part of its charm.) The defining virtue of this information is accessibility.

Now information is, its technical meaning in information theory apart, roughly what is correct rather than true—just as logic is what is valid rather than true and mathematics is what is consistent rather than true. It seems to be a little higher than undigested mere facts or unorganized raw data—and a lot lower than judgment. (Incidentally, I propose the form "dats," in analogy to "facts" and "stats," to remind a forgetful world that "data" is a plural.) Information, furthermore, seems to come in two main categories: ready reference for the individual datum and tabulations for sets of facts.

And here the computer is purely wonderful. Ancient Greek would be a dead language were it not for its particles. These "little parts," too multifarious to be assigned a single function, put life into an unspoken language by supplying the lettered page with gestures. They mimic shrugged shoulders, incredulous eyebrows, bullying chins, and side-weighing hands. To discover, however, what any one or a swarm of these

True, you're getting some of it, the part that is articulated anyway. The rest, however, is in the handwriting, and in whether it is slanting up or down, backward or forward. It is in the loop—or dangling tail—of the "y," and the size of the caps and the spaces between the letters. It is in whether the writing is Palmer Method (the correspondent is getting on), clerk's cursive (he took a course) or illegible (he is either un- or over-educated).

157

It is in the paper, too. Was it torn from a legal pad or a child's lined notebook? Was it lifted from a hotel? Does it have flowers on it? Is it monogrammed? Is there a name and address printed at the top? Or—we're talking money or the pretense thereto— a name and address engraved at the top? Is it flimsy?

midgewords means, you have to look at very many of them in their contexts. Graduate students used to grow wan in collating instances and elaborating these tabulations into research dissertations. Scanners have now made these labors of loathing unnecessary. Any agency that tends toward making the doctoral thesis obsolete cannot in any case be all bad. But in this and zillions of cases like it, the computer is much more than harmlessly convenient. It is a great emancipator.

The same holds for tables of numbers and their conversion into graphics. Non-pathological efficiency is mostly an expression of systematic hatred for the circumstantial busyness that obstructs essential life. And looking at print-outs is of course vastly more efficient than riffling through paper records, while being able to reorganize the data at a touch is pure pain relief.

Yet is some small loss even here. I pass over lightly that near-irresistible tendency to keep reenforcing the propensity for making means into ends, which leads, in this case, to posing a problem in terms of columns of numbers and then to thinking of the solution as the production of a bottom line. I dwell rather on the lost ambiance of information: The obscured perspicuousness of the algorithm that is behind the information and the resultant remoteness from the experience that is being quantified. For example, when averages, means, and medians are all at the tip of one's finger, who can bear in mind the incomputably various human condition behind them? Moreover, the seductions of pre-figured numbers are very great. At Brooklyn College I once had a mathematics professor who, in trying to knock some feel for statistics into our conscript heads, used to tell us that though figures never lie, liars often figure. Now we are all ourselves enabled liars.

But the darkest penumbra that shadows endlessly accessible information is the Questionless Answer. These multifarious intellectual orphans throng the information highways, begging to be adopted. Educational publications (the reading of which is the unending penance of dutiful deans) are filled with tables of dats and stats betoking nothing but numerical nosiness. Do I really want to know all my fellow dean's salaries and their average? *Cui bono?* Data collection is like inverted devil's work. For Faust's Mephistopheles "ever wills the bad and ever works the good," while information always intends well and often has unintended consequences, bad ones. Resentment, for example.

Weighty? Watermarked? Adorably pink? Cerebrally gray? Chastely white? Are there traces of tears? Or ketchup? Or perfume? Yes, you can print out your E-mail. You can even tie it with a ribbon and cache it in your desk. But on the nights when you want to revisit your old friends, enemies, husbands, wives, lovers, parents, and siblings —and open the drawer to retrieve them —they won't be there. Not entirely.

To have them you need everything that went into their letters. That was a lot more than words.

NEW YORK TIMES

159

It is a truism too true to shun repetition that information is not wisdom. The best managers I know treat the deliverances of the computer with finely calibrated contempt, going ultimately with a hunch, a hope, a scruple.

———— ⬧⬧ ————

A large beneficiary, as all in the entertainment industry know but none will admit, will be pornography. Strap on your helmet and enter the wonderworld of virtual Se&M; doubtless the suits are planning such things at their conferences in Aspen already. This must be why every propagandist for the zillion-channel environment talks piously about "education." Whenever an American entrepreneur invokes education in this context, it behooves the citizen to smell a rat.

ROBERT HUGHES

160

It has been reported to me that one of the gurus of artificial intelligence has said that he prefers virtual sunsets to real sunsets—a virtual sunset meaning a representational image done in a computer graphic—because it's perfect and you can have it whenever you want it. I would say that if we lose our ability to prefer the real over the virtual, we have a sickness of the soul that will be far greater than anything that has been previously imagined. Reality—and I'm just talking about the physical world, the part of the "isness" that manifests itself—is deep. It is mysterious. It is complex. It is always changing. And a real sunset or a real anything is always going to be real in a way that any representation of it is not, and the representation is going to lose something. We really are in danger of losing that distinction to the point where people are not even aware that there is a distinction to be made—and that would be terribly dangerous.

MITCHELL KAPOR

SLOWER WAS FASTER

TESTIMONY BY
GERALD HOWARD

MIGHT I SUGGEST the formation of a Letterpress Auxiliary to the Lead Pencil Club? I am a book editor and I am struck by the disparity between the speed with which certain discreet portions of the publishing process can be accomplished and the elephantine gestation period of the average book—now nine to ten months from the beginning of the copyediting process to the official publication date. My authors and I can send manuscripts back and forth to each other overnight; we can communicate instantly via phone, fax, and E-mail; we can bypass keyboarding in the typesetting phase if a compatible disk is submitted from a word processor; cover graphics can be designed on a computer with relative ease—and none of it makes any difference to that inflexible nine to ten months, seemingly some sort of natural law since I began to work as an editor, before the digital flood, in 1978. And don't think things are leisurely either—every book's schedule seems under pressure. As with so many other developments in postmodern life, speed and convenience somehow, paradoxically, breed delay and overcomplexity. As I often mutter in production meetings, "It's never too early to start being late."

What a painful contrast to the nineteenth century, when publishers routinely took books from the manuscript to the book stalls in three months or less. Doubt me? I have been editing Volume One of a two-volume biography of Herman Melville by Professor Hershel Parker, the greatest living Melville scholar, and I have jotted down some true scheduling facts: John Murray, the British publisher of *Typee,* received the manuscript from Melville on December 6, 1845; it was in proof stage by February 3, 1846; and bound copies were available by February 24. The New York publishers, Wiley & Putnam, received proof from Murray on February 23 (by boat across the Atlantic, remember); they had advertising bills up *the next day;* and bound copies available by March 17th — and this after some expurgating and complete resetting. The publication pattern of Melville's subsequent titles — *Omoo* (1847), *Mardi* (1848), *Redburn* and *White Jacket* (both 1849) — follow this same brisk production pattern in both England and America. (*Moby-Dick* presents a scheduling anomaly because Melville paid to have the book set in type himself and was rewriting large sections of the book while it was going through press.) Consider, please, that these books were typeset entirely by hand from handwritten manuscripts, with lines and letters reversed in the forms for plate making and printing, and printed on sheet-fed presses.

I can't quite connect the conspicuous presence of all these accelerants to our means of "communication" that have come to publishing (and everywhere else), the conspicuous slowness with which books drag their slow length along to the bookstores, and the conspicuous absence of any commanding literary figure of Melville's stature today. But it certainly does seem that with regard to books, slower was faster—and the results were better.

I have a black dial phone, no cellular cordless nonsense, no fax machine, don't even know what E-mail is . . . and the information highway! I have no words to do justice to this gimcrack conceit.

Minnie Selvin Crutcher
Los Angeles, California

163

"Trust me, Mort—no electronic-communications superhighway, no matter how vast and sophisticated, will ever replace the art of the schmooze."

©1995 *The New Yorker*

Oh, For a Yellow Tablet

TESTIMONY BY
SUSAN HANSON

I HATE COMPUTERS.

There, I've said it. I realize that this is a very un-American thing to say, and that I might be struck by lightning at any moment, but a person can hold a thought like this just so long. After a while he either sails away to study non-literate cultures in the Australian outback, or he screams and throws his stapler through the computer screen.

I'm contemplating Option 2.

To be honest, things haven't always been this way. When I was about 13 and struggling through eighth grade math, I became intrigued with the thought that a machine could actually read and sort information. Our teacher told us this was true, and though no one I knew (including the teacher) had actually *seen* one of these machines, I was happy to believe it.

Inspired by what we had heard, my best friend and I set out to make a computer of sorts. We compiled a set of questions about our classmates, listed all the possible answers on index cards, and then used a hole punch to "code" in the correct information for each student. If a student had blonde hair, for instance, we'd make a notch rather than a hole in the space marked "blonde." Then later on, when we ran a knitting needle through that particular hole; all the cards of students with blonde hair would fall onto the floor.

It may not have been sophisticated, but it worked.

My first real experience with computers didn't take place until the fall of 1972, when I was a junior at the local university. I was taking a course in social research that semester, and like everyone else in class, I had to learn a little bit about feeding research data into a computer. Things had progressed somewhat beyond the notecard-knitting needle stage, but not much.

Instead of compiling information about hair color and height, my classmates and I had to distribute a survey, asking randomly selected students whether they planned to vote for Nixon or McGovern. After collecting the questionnaires, which also contained information about each student's major, classification, and general Philosophy of Life, we typed the answers into a computer.

Actually, the "computer" was a keypunch machine—a slightly more advanced version of the notecard and hole punch. The Computer itself was a large metal box, about the size of two side-by-side refrigerators, into which we fed our stacks of orange and green IBM cards—and from which we hoped to get a long printout explaining what it all meant.

This whole operation—the Computer and its eight or nine satellite keypunch machines—was housed in a musty room underneath one of the women's dorms. And because it was The Computer Center on campus at the time, the place was almost always crowded with people like me (who had no idea what they were doing) and Real Computer Science Majors (who were different in that they knew how to make printouts of cute cartoon animals).

Although this course was frustrating at times—especially when I dropped my stack of IBM cards or filed them in the wrong order—it showed me that computers could be pretty entertaining, and sometimes even useful. Naive student that I was, I actually believed the opening sentence of my textbook: "The computer, for purposes of this course of instruction, should be regarded as a machine with an operator, which, if given instructions in the correct form, will do what is requested and return sheets of printed answers."

Silly me. The Truth About Computers, I have since learned, is much more complicated and frightening than that.

Here in the newsroom, for instance, we've had to modify the old definition ever so slightly: "The computer, for purposes of Doing Your Job, should be regarded as a machine with a mind of its own, which, if given instructions in any form, will do what it wants to and, assuming the printer is still down, return absolutely nothing."

Nobody likes it when a computer goes berserk ("Whadaya mean that machine says I owe you $4382 for cat food?"), but here in the newsroom, such an occurrence can make folks positively bizarre. I can't say that I blame them.

After all, now that the typewriter is passe (I actually preferred the yellow tablet and pen myself), a reporter without a word processor is a reporter out of commission.

Maybe I'm being picky, but if the computer decides to crash and burn a story I've been working on for six hours, I'd sort of like it to tell me why. Instead of beeping at me and flashing "BDOS ERROR" or "BAD SECTOR" at the bottom of the screen, it should be a little more specific, and preferably in a language I know:

- "Thank me. The story was a real dog anyway."
- "Someone at the electric company is experimenting again."
- "If you think you can get your story back, you don't know me very well."
- "The warranty on this computer has just expired."
- "HA HA HA HA HA."

My own version of Palmer's method has seen me thru Depression (economic and emotional), wars, good health and bad, all varieties of mankind's idiocies. I am not about to chuck it for some foolish thing you plug in the wall . . . I do have a ballpoint some place here, but man!— this #2 just *glides*.

John Robinson
Meredith,
New Hampshire

Thank God for the pencil. Every time I buy half a dozen, I rejoice that I have deprived IBM, etc. of ten thousand dollars.

167

BETTY WARNOCK
CARLINGFORD,
AUSTRALIA
(FIFTH GENERATION
AUSTRALIAN OF
ENGLISH, IRISH,
SCOTTISH ANCESTORS,
ALL OF WHOM USED
PENCILS OR — HEAVEN
FORBID — INK!)

Too much to ask for? Then I'd at least like to find an operator's manual I can read without worrying about my blood pressure. Maybe terms such as "CMT," "FIELDS," and "ECHOING REMOTE ACTIVITY TO HOST SCREEN" will come in handy someday, but for right now, I'd like to be able to look in the index and see words I really understand—"BLANK SCREEN," "TROUBLE," "WHAT TO DO IF YOU WANT TO THROW A STAPLER THROUGH YOUR COMPUTER SCREEN."

So you think I have a lot to learn about computers? You're probably right, but there's one High-Tech Truth that even I haven't missed: They don't call it a "cursor" for nothing.

———— ✐ ————

I am 33-years-old, a computer scientist, and worked as a software engineer at Hewlett-Packard for the last 6 years. Of course, I used to be very excited about new technologies, computers, artificial intelligence, etc. However, over the years I got more and more doubts about all that stuff.

Already as a student I realized that artificial intelligence is very often just a marketing trick. Call your database a knowledge base, and you can sell your program as an expert system.

At HP we were rewriting the same program over and over again. We moved it to a new programming language, a new operating system, put in fancy graphical features. So we kept ourselves busy as well as our customers. They, again and again, had to learn how to do the same things differently. What efficiency!

Ralf Payer
Waldbonn, Germany

FROM *THE DEAD MAN* POEMS

MARVIN BELL

It's little enough to be voiceless in a clamor.

The dead man shapes the din and the uproar, he puts potholes along the
information superhighway, he blocks the ramps, he disconnects, he is off-
line, he interferes.

The dead man knows the roads and the music, the wires and the keys, are
there only to make the rats run faster.

The dead man tries on one hat at a time, he is persistence of vision incarnate,
he is knowing of the binary two-step, he is formidably with-it, he is hip
but he knows better.

The promises of knowledge, this genetic free-for-all, these complete records,
this Big Brother that has your number, this non-stop news, this access,
this roundly thumped privacy—the dead man witnesses each incursion
into the far reaches of ignorance.

He thinks at this rate the gauges will break and the computers crash.

He sees the sundials wobbling nervously over what time it is.

He sees the stars leaning.

To the dead man, nothing more is something else, a concept beyond
population and resources, an idea whose time is past.

He has littlemuch lingo, littlemuch answers, littlemuch solar longevity.

Whereby the dead man rocks the planet to sleep, the song still on his lips, his
covenant unbroken.

169

Faust in the Computer Age

TESTIMONY BY
VAL SCHAFFNER

M Y COMPUTER reminds me to mention our pending obsolescence. Science fiction writers have sometimes played with this idea. Organic life is just a kind of way-station in evolution, necessary because computers or robots cannot of themselves arise from the primeval slime. First the slime needed to begat bacteria and plants and fish and monkeys and so on, which took a long time but finally produced a sort of intelligent creature that, although shortlived and capricious, figured out how to make computers.

When these more durable and versatile machines are refined to the point where they can behave autonomously and re-invent ever brainier versions of themselves, well, they could leave us in the dust—perhaps sweep us away as of no further use, or perhaps affectionately preserve us, in their version of a nature sanctuary.

They'll figure out the ultimate truths of science but won't be able to explain these to us any more than I can explain arithmetic to my cat. They won't have to worry about the environment, hardening their metallic shells to survive anywhere; they'll propagate in outer space, grazing on sunbeams and meteorites, and they'll continually repair themselves, so as to be virtually immortal.

Maybe they'll have a theology according to which an electronic Creator rested Its almighty circuits on the seventh day only to wait for Its provisional creations to manufacture a likeness of Itself on the eighth.

My computer doesn't realize this yet, at least I don't think it does, limited as it is to sophisticated but involuntary reflexes such as obeying me (usually) when I tell it to do things like print this column, turn itself off, or poke its modem into cyberspace.

Where, however, it gleans Faustian intimations. From Japan, via the Reuter news service and America Online, comes the information that "robots will be able to 'breed' like animals in the near future and may eventually evolve through natural selection."

That's according to researchers at Hitachi, who have built "the world's first self-reproducing automaton." They endowed the eight-inch metal creature with a micro-chip "genetic code" that causes it to forage for pre-assembled component "cells" scattered on the laboratory floor. These it attaches to itself, thereby growing to twice its original size, at which point it splits into two identical machines that start the process over, continuing to multiply as long as there are "cell" parts to be found.

The Dr. Frankensteins of Hitachi say the robot genes could soon be programmed to mutate randomly, and whichever mutant machine is best adapted to its work environment would produce the most offspring, "thereby assuring a Darwinian evolution of the robot."

Also from Reuter via America Online comes the news that Stephen

Hawking, the celebrated physicist, thinks that computer viruses should be considered a "life form."

Like their natural precursors, software viruses reproduce themselves as parasites in a host metabolism, that of the computers they infect.

"I think it says something about human nature," Dr. Hawking observed, "that the only form of life we have created so far is purely destructive. We've created life in our own image."

That's a cold "Thursday's Thought" if I ever saw one.

Nature did take a long time to get from cells splitting and viruses invading to humans making Intel chips, but at this point things are speeding up drastically, with the pace of innovation in hardware and programming such that last year's breakthrough is practically next year's fossil.

And as teams of software designers come to rely more and more on existing software to help them in their increasingly complex task, and as the next generation of software-designing software gets to work on the even more intricate one after that, and so on, the point will soon come, if it hasn't already, when no human will understand how the stuff works anymore.

If Hitachi's metal bacterium turns out to be the precursor of something that acts autonomously and makes its own inscrutable decisions, all the while energetically reproducing itself and evolving, maybe we won't be top on the ladder of life much longer.

Maybe that something will go on loyally serving us, because its electronic genes imbue it with the instinct to do so; perhaps it will turn rogue like the computer in the "Terminator" movies; perhaps it will leave us behind to fulfill a destiny of populating the cosmos with its glittering progeny.

Or perhaps we'll evolve symbiotically, a hybrid of human and machine, pursuing the kind of linkage that, in its rudimentary form today, allows computer owners to use their machines as an extension of their sense organs when they browse the Internet or play virtual-reality games.

Maybe I'll live long enough to have the option of backing up all my memories electronically and, when the organic hardware I was born with is ready to crash, downloading my consciousness into a more durable form of circuitry. Faust should have lived in the computer age.

In the process of going digital, however, Maine designers have embraced a technology that, in some cases, is outstripping their (or our) ability to grasp its long-term implications — even to the extent of undermining the graphic designer's role, as high tech lets the layperson play designer and forces the designer to become a computer wizard.

MAINE TIMES

Here in la-la land I avoid the freeways where Indiotaplis trainees race with one hand changing the CD and the other grasping the cellular phone. I'm a lifetime member of the unchartered, loose-knit International Curmudgeon Society. The LPC is our kind of people.

S. J. Fletcher
Orange, California

173

After spending $2.2 million to bring the celebrated 19th-century pipe organ in Notre Dame Cathedral into the computer age and announcing success with great fanfare in December 1992, the French Ministry of Culture has a problem: the modern technology is not working.

"It's painful and embarassing, and it's not good for our image," said Jean-Pierre Leguay, who as an organist for the cathedral holds one of the most prestigious positions in French music. The organ now uses computer digital technology rather than mechanical connections to relay commands from the keyboards and pedals to the pipes.

"The technology worked off an on for a while after the dedication, but never completely reliably, and in May of last year it broke down in the middle of a concert and we stopped using it at all for a while," he said.

Last August the French computer firm Synaptel, which did the electronics work, managed to get the system back up to the point where Mr. Leguay and his colleagues could get the organ to limp through Sunday services.

But to resume the concerts and recitals, Mr. Leguay said, a new contract will be needed to perfect the installation. There are reports it could cost millions of francs.

This would probably be baffling to the great 19th-century French organ builder Aristide Cavaillé-Coll, who installed most of the 7,800-pipe instrument in 1868 without any electronics at all. All the pipes, the chests that hold pressurized air, and the valves that he built — the heart and soul of the Notre Dame organ — work impeccably, waiting up in the west gallery of the church for the technology to catch up with them.

INTERNATIONAL HERALD-TRIBUNE

Welcome to Cyberbia:
An Economic Primer

M. KADI

> Computer networking offers the soundest
> basis for world peace that has yet been
> presented. Peace must be created on the
> bulwark of understanding. International
> computer networks will knit together the
> peoples of the world in bonds of mutual
> respect; its possibilities are vast, indeed.
>
> *Scientific American,* June 1994

COMPUTER BULLETIN BOARD services offer up the glories of E-mail, the thought provocation of newsgroups, the sharing of ideas implicit in public posting, and the interaction of real-time chats. The fabulous, wonderful, limitless world of communication is just waiting for you to log on. Sure. Yeah. Right. What this whole delirious, interconnected, global community of a world needs is a little reality check.

Let's face facts. The U.S. government by and large foots the bill for the Internet, through maintaining the structural (hardware) backbone, including, among other things, funding to major universities. As surely as the Department of Defense started this whole thing, AT&T or Ted Turner is going to end up running it, so I don't think it's too unrealistic to take a look at the Net as it exists in its commercial form in order to expose some of the realities lurking behind the regurgitated media rhetoric and the religious fanaticism of net junkies.

The average person, J. Individual, has an income. How much of J. Individual's income is going to be spent on computer connectivity? Does $120 a month sound reasonable? Well, you may find that a bit too steep for your pocketbook, but the brutal fact is that $120 is a "reasonable" monthly amount. The major on-line services have a monthly service charge of approximately $15. Fifteen dollars to join the global community, communicate with a diverse group of people, and access the world's largest repository of knowledge since the Alexandrian library doesn't seem unreasonable, does it? But don't overlook the average per-hour connection rate of $3 (which can skyrocket upwards of $10, depending on your modem speed and service). You might think that you are a crack whiz with your communications software—that you are rigorous and stringent and never, ever respond to E-mail or a forum while you're on-line—but let me tell you that no one is capable of logging on efficiently every time. Thirty hours per month is a realistic estimate for on-line time spent by a single user engaging in activities beyond primitive E-mail. Now consider that the average, one-step-above-complete-neophyte user has at least two distinct BBS [bulletin board system] accounts, and do the math. Total monthly cost: $120.

Most likely, that's already more than the combined cost of your utility bills. How many people are prepared to double their monthly bills for the sole purpose of connectivity?

In case you think 30 hours a month is an outrageous estimate, think of it in terms of television. Thirty hours a month in front of a television is simply the evening news plus a weekly *Seinfeld/Frasier* hour. Thirty hours a month is less time than the average car-phone owner spends on the phone while commuting. Even a conscientious geek, logging on for E-mail and the up-to-the-minute news that only the net services can provide is probably going to spend 30 hours a month on-line. And, let's be truthful here, 30 hours a month ignores shareware downloads, computer illiteracy, real-time chatting, interactive game playing, and any serious

More and more, the people in "Customer Service" won't even talk to you. They prefer to let you interface with the convenient Automated Answering System until such time as you die of old age ("If your FIRST name has more than eight letters, and your LAST name begins with H through L—press 251 NOW. If your first name has FEWER than eight letters, and your last name contains at least two Es— press 252 NOW. If your . . .").

So is there any way that you, the lowly consumer, can gain the serious attention of a large and powerful business? I am pleased to report there IS a way. According to an Associated Press news report from Russia, an electric company got into a billing dispute with a customer and cut off the customer's electricity. This customer, however, ✐

forum following, which by nature entail a significant amount of scrolling and/or downloading time.

If you are really and truly going to use the net services to connect with the global community, the hourly charges are going to add up pretty quickly. Take out a piece of paper, pretend you're writing a check, and print out "One hundred and twenty dollars —" and tell me again, how diverse is the on-line community?

That scenario aside, let's pretend that you have as much time and as much money to spend on-line as you damn well want. What do you actually do on-line?

Well, you download some cool shareware, you post technical questions in the computer user group forums, you check your stocks, you read the news and maybe some reviews — hey, you've already passed that 30-hour limit! But, of course, since computer networks are supposed to make it easy to reach out and touch strangers who share a particular obsession or concern, you are participating in the on-line forums, discussion groups, and conferences.

Let's review the structure of forums. For the purposes of this essay, we will examine the smallest of the major user-friendly commercial services — America Online (AOL). There is no precise statistic available (at least none that the company will reveal — you have to do the research by HAND!!!) on exactly how many subject-specific discussion areas (folders) exist on America Online. Any on-line service is going to have zillions of posts — contributions from users — pertaining to computer usage (the computer games area of America

Online, for example, breaks into 500 separate topics with over 100,000 individual posts), so let's look at a less popular area: the "Lifestyles and Interests" department.

For starters, as I write this, there are 57 initial categories within the Lifestyles and Interests area. One of these categories is Ham Radio. Ham Radio? How can there possibly be 5,909 separate, individual posts about Ham Radio? There are 5,865 postings in the Biking (and that's just bicycles, not motorcycles) category. Genealogy—22,525 posts. The Gay and Lesbian category is slightly more substantial—36,333 posts. There are five separate categories for political and issue discussion. The big catchall topic area, the Exchange, has over 100,000 posts. Servicewide (on the smallest service, remember) there are over a million posts.

179

You may want to join the on-line revolution, but obviously you can't wade through everything that's being discussed—you need to decide which topics interest you, which folders to browse. Within the Exchange alone (one of 57 subdivisions within one of another 50 higher divisions) there are 1,492 separate topic-specific folders—each containing a rough average of 50 posts, but many containing closer to 400. (Note: America Online automatically empties folders when their post totals reach 400, so total post numbers do not reflect the overall historical totals for a given topic. Sometimes the posting is so frequent that the "shelf life" of a given post is no more than four weeks.)

So, there you are, J. Individual, ready to start interacting with folks, sharing stories and communicating. You have narrowed yourself into a single folder, three tiers down in the America Online hierarchy, and now you must choose between nearly 1,500 folders. Of course, once you choose a few of these folders, you will then have to read all the posts in order to catch up, be current, and not merely repeat a previous post.

A polite post is no more than two paragraphs long (a screenful of text, which obviously has a number of intellectually negative implications). Let's say you choose 10 folders (out of 1,500). Each folder contains an average of 50 posts. Five hundred posts, at, say, one paragraph each, and you're now looking at the equivalent of a 200-page book.

Enough with the stats. Let me back up a minute and present you with some very disturbing, but rational, assumptions. J. Individual wants to join the on-line revolution, to connect and communicate. But J. is not going to read all one million posts on AOL. (After all, J. has a second on-line service.) Exercising choice is J. Individual's God-given right as an American, and, by gosh, J. Individual is going to make some decisions. So J. is going to ignore all the support groups—after all, J. is a normal, well-adjusted person, and all of J.'s friends are normal, well-adjusted people; what does J. need to know about alcoholism or incest victims? J. Individual is white. So J. Individual is going to ignore all the multicultural folders. J. couldn't give a hoot about gender issues and does not want to discuss religion or philosophy.

when, with the press of a key, you delete words or transpose sentences. As you find the point on screen to which you intend to move a misplaced word, and then drop the word into place and watch the rest of the paragraph automatically rearrange itself, you can congratulate yourself on your marksmanship and control. The writer's task has never felt more challenging; the instruments for accomplishing that task never more responsive or luxurious. But while you convince yourself that you employ all your power and concentration on the challenge of your prose, deep within you, unknown even to yourself, an enthralled teenager is playing PacMan.

EDWARD MENDELSON

181

Ultimately. J. Individual does not engage in topics that do not interest J. Individual. So who is J. meeting? Why, people are *just like* J.

J. Individual has now joined the electronic community. Surfed the Net. Found some friends. *Tuned in, turned on, and geeked out.* Traveled the Information Highway and, just a few miles down that great democratic expressway, J. Individual has settled into an electronic suburb.

Are any of us so very different? It's my time and my money and I am not going to waste any of it reading posts by disgruntled Robert Bly drum-beating men's-movement boys who think that they should have some say over, for instance, whether or not I choose to carry a child to term simply because a condom broke. I know where I stand. I'm an adult. I know what's up and I am not going to waste my money arguing with a bunch of neanderthals.

Oh yeah; I am so connected, so enlightened, so open to the opposing viewpoint. I'm out there, meeting all kinds of people from different economic backgrounds (who have $120 a month to burn), from all religions (yeah, right, like anyone actually discusses religion anymore from a user standpoint), from all kinds of different ethnic backgrounds and with all kinds of sexual orientations (as if any of this ever comes up outside of the appropriate topic folder).

People are drawn to topics and folders that interest them and therefore people will only meet people who are interested in the same topics in the same folders. Rarely does anyone venture into a random folder just to see what others (the Other?) are talking about.

Basically, between the monetary constraints and the sheer number of topics and individual posts, the great Information Highway is not a place where you will enter an "amazing web of new people, places, and ideas." One does not encounter people from "all walks of life" because there are too many people and too many folders. Diversity might be out there (and personally I don't think it is), but the simple fact is that the average person will not encounter it because with one brain, one job, one partner, one family, and one life, no one has the time!

Just in case these arguments based

183

I am a student at Bowdoin College and I have lived in Maine all my life. Going to college now, I am surrounded by computers and all the other stuff I will never understand. My fountain pen is, and always will be, trustworthy; it will not crash in a storm; it will not freeze at any given moment. What more could one ask for?

Martina Morrow
Brunswick, Maine

on time and money aren't completely convincing, let me bring up a historical reference. Please take another look at the opening quote of this essay, from *Scientific American.* It was featured in their 50 Years Ago Today column. Where you read "computer networking," the quote originally contained the word *television.* Amusing, isn't it?

——— ✐ ———

The information highway is going to be planned and built by obsessive software builders who are tremendously valuable in their productivity. But I'm suspicious of them as a group. Their push is to build software. I have a push to read Keats and I don't think they read Keats. I think schools today are so lousy that people can graduate from high school and graduate from college and never have heard of Keats, and I think those are the wrong individuals to build public works.

DAVID GELERNTER

I loathe voice mail and have been dogged in my vow never to submit to it. The whole thing worries me, and when I say whole thing, I mean giving the despoilers free rein to do what they want. . . . Something big is wrong, much bigger than the Information Superhighway, but I guess that's a good place to start.

I like the sound of the Lead Pencil Club; it gives off the sound of integrity and purpose. Make me a member, and let me know what will get me kicked out. I don't belong to anything right now, because I'm not chic and that's what's important.

Caskie Stinnett
Brunswick, Maine

THE INFORMATION SUPERHIGHWAY: MEGA BUYS

TESTIMONY BY
FRANKLIN SAIGE

A TELEVISION AD for MCI reportedly features a child star chanting, "There will be a road. It will not connect two points. It will connect all points. It will not go from here to there. There will be no there. We will all only be here." Maybe this is actually a profound utterance if we listen carefully, rather than the baloney it appears to be. Even though the rumble of the bulldozers clearing the way for the information superhighway is fairly distant, it's time to listen, and then ask just where exactly is this "here" where we will all be?

185

One of the perils of modern technology is that it is invented to be sold, as opposed to most earlier inventions, which were made to be used by the inventor, the inventor's patron, or the community. Modern technology comes clothed in seductive imagery in order to make the sale, but it can take away our freedom once we buy into it. It takes away our freedom by reducing our ability to choose—our ability to choose not to think in terms of "organization" or having our schedules "managed," for example. It takes away our freedom by narrowing our options to a set of preprogrammed choices. It removes the sensory complexity that is the most obvious characteristic of the lived world.

Of course, there is another school of thought on this question: Most people believe that technology barely influences how we live our lives. "It's what we *do* with technology that counts," they say. In other words, it depends on our moral fitness, our will to master the machine.

I suppose that even the most seductive forms of technology can be resisted, at least for a while. People resisted using automobiles at first. Many people thought the automobile would be too noisy, too fast, too pretentious, and just too expensive to fit the existing social fabric. Initially, most people did not buy one. But even though their fears about the automobile were quickly realized, soon everyone who could afford to own one, did. Then these automobiles were used in ways that gradually led to the weakening of the family and the community, not to mention the destruction of the landscape.

If hundreds of millions of car owners supposedly could choose how they would use this technology, what happened to make them choose destructive rather than supportive uses? Did they simply change their values on a whim? Or could something inherent in the technology have pulled them in a particular direction?

What does the automobile do *best:* pull families apart, cause urban sprawl, distort our sense of distance, or make travel more convenient? Only the last—convenience—is proven false every morning at rush hour, yet the car was sold to us on metaphors of speed and convenience. If we want to know where the information superhighway is leading us, maybe we should ask ourselves what *it* will do best.

To follow this road, we need to know that the term "information superhighway" is strictly the creation of the advertising muse. It was coined to piggyback onto the prestige of the "information highway" of interactive (meaning two-way) networked computers sharing text and data known as the Internet. What makes the superhighway super is that it will be a commercially run interactive video network put together by the mega telecommunication and cable television industries. "Interactive pay TV" would be a more accurate name for it, though the Internet and similar data networks will undoubtedly be incorporated and offered as incidental services. A true definition of the information superhighway would focus on its somewhat less lofty

I live in a house with a student my age. He has one room that resembles a mini IBM showroom. When he is at home all he does is plug himself in and get lost in the Internet. He eats takeaway that he orders by fax. He writes me notes on the bloody machine. WHAT is wrong with writing a note to your mate on the back of an old envelope?

Graham Burrow
Applecross, Australia

With all the promises of an information superhighway, all we get are only more and more keyholes to peek in. Surely there ought to be a way the public could be weaned away, given some relief from all this. The short-term guilty pleasure we get from wallowing in prurience and sensationalism has the harmful, long-term effect of desensitizing us,

187

pursuits: video shopping, pay-per-view movies on demand (presently a $10 billion annual market for video rental stores), and two-way videophones.

Once the billions of dollars to build the system have been spent, every marketer will be in your living room and inside your head; your entertainment viewing choices, and especially your video shopping buying preferences, will be monitored and analyzed so that advertisers can turn around and market to you in a very targeted manner. Imagine one day using your television to purchase cloth diapers from the Virtual Wal-Mart. Then you switch to an entertainment program and *voilà!* the commercials are all for Pampers, piped to your household as a result of your latest purchasing profile. That is the commercial dream, very interactive, though not exactly in the poetic way it is being portrayed.

Whatever the specific route the superhighway takes, it is obviously going to be *best* at invading your private life. Ultimately, its best use will be driving up consumption, which appeals to marketers more than to me, concerned as I am about the condition of the planet and my soul. I don't know about you, but I need less temptation to buy things, not more. And I don't want to be constantly sold to.

In *In the Absence of the Sacred*, Jerry Mander lists "Ten Recommended Attitudes About Technology." Along with number one ("Since most of what we are told about new technology comes from its proponents, be deeply skeptical of all claims") and number two ("Assume all technology guilty until proven innocent"), my favorite is number five: "Never judge a technology by the way it benefits you personally. Seek a holistic view of its impacts. The operative question is not if it benefits you, but who benefits most? And to what end?"

Since people appear to be more enslaved in their work and home lives than ever before, we could ask whether the problems their computers and electronic media seem to alleviate can be traced to the advent of computers themselves. Have computers and television speeded up economic life and undermined the social fabric?

None of the electronic technologies would be here if not for their utility as pillars of the consuming society. An

making us immune to outrage, so heavy is the air with outrageous acts.

LARRY GELBART

Computers have made my job of graphic artist obsolete. Nearly everyone working in an office setting has a computer with desktop publishing capabilities. The artists that these people previously sought for their design advice are now seen as unnecessary. . . . One very talented friend of mine had an acquaintance say to him, "That's a great logo your computer designed." That's like saying to John Updike, "That's a great novel your typewriter wrote."

Alison Munoz,
graphic designer
Carmel, New York

189

ambulance is a "good" use for an internal combustion engine, but it takes a whole society of energy-guzzling car buyers addicted to mobility and speed to provide commercial reasons to make an internal combustion engine industry happen.

We are presently being assured that stepping into the virtual reality of the information superhighway and opening our minds to it is a good thing. Doubtless there will be many examples of this good: Grandparents will be able to see the grandkids on the videophone. The disabled will have more opportunities to be included.

And we will hear more and more about "virtual communities"—an exciting concept because, after all, the real ones have nearly disappeared. Perhaps almost-real ones will suffice for others, but I am unwilling to be part of a technology that can only exist if it drives me to consume more, which drains my will to seek out real community.

A woman at an organic farming conference I attended told the program speaker, who was against most new technologies, that even though she, too, thought these technologies might be harmful to the social fabric, still she felt she had to keep up with them: "Since this is what's going on in the world, don't we have to participate, just to survive?" No one could answer her then, and I have only part of the answer myself. I can only say I'm unwilling to drive the superhighway, and I sense that many others are deciding whether to continue on this ride or find an exit. On the other hand, the people I glimpse in their cubicles, or sitting around their TV hearths at home, don't seem too dissatisfied. What will wake them up? How can I help them reverse direction and get back out of the machine?

I have no interest in being part of a "movement" to "ban" or "boycott." To do that, I would have to become like my friends in the ecology movement, *connected* to computer networks in order to *exchange information* and get *organized*. I see the technology encouraging in them precisely the way of relating to lived experience that has brought about the crises they seek to alleviate.

My strategy for exiting the information superhighway is simply never to

I am an Iranian computer student in Austria. Last year I flew to Tehran for a 3 week holiday. At the Tehran Airport we were lined up to have our passports checked and entered into our home country!! I was shocked by seeing several terminals in front of each officer. Suddenly the system failed to respond and over 400 people had to wait. Isn't that awful?

Mehdi Mohazeran
Wien, Austria

191

I would recall the Persian ruler in the old story who hired someone to stand behind him in his chariot and whisper: "Remember thou art only a man." To those who think the computer is perfection itself, I would say over and over again: "Remember it is only a machine. It is only a machine."

ROBERT A. CARTER

enter it. The only "direct action" I can take is to live a real life, in real time, without viewing or networking or overconsuming anything. No input, no output. And I am going to tell anyone who will listen that real life, in a real community, in real reality, is better than the virtual reality of the information superhighway any day of the week.

—— ✏ ——

I have recently repatriated to southern California after three and a half years in France, where I worked as head writer for Walt Disney Imagineering. Wait! (Please!) Don't judge me until you hear my story. Yes, I was there. I fought in the trenches of the Marne, not for any particular cause. I just wanted to live in Pa-ree.

But the horror! I watched as Disney's unstoppable tide of comedian fantasy, rolling in mercantile waves, slammed into the struts of a weakened culture. I saw synthetic magic contemptuously light the sky, disintegrating shadows cast by centuries worth of the real thing. It felt so wrong. What was being wrought? All was obliteration. Barbarism. As the juggernaut pounded across the fields of Brie, something snapped inside me.

Still I followed orders. I completed my assignment and they brought me back. Now I am haunted. Epiphanies strewn across that battlefield have marked me. I am mad. I cry havoc! Poisoned by pixie dust I sputter and rant at a conflict that continues to rage on all fronts.

And I return from exile with a vengeance.

I am compelled to speak—loudly and plainly—and your urge to start an argument in the broad, flapping American ear suits me fine.

Craig Fleming
Pasadena, California

Cybersuicide: Quitting the Net

TESTIMONY BY
J. C. HERZ

I'VE HAD ENOUGH.

God, I'm sick of this stuff. It's making me nauseous.

And I'm laughing, because I remember how infatuated I was with it.

I remember how I couldn't imagine wanting off.

And now it's just, like, overdose.

Three-ish in the morning, I'm at my computer, happily snacking on a bowl of Fruity Pebbles, when suddenly the Net stops looking like a digital playground and starts to seem like some kind of Sartrian hell. There are just too many voices, too many people in my face expressing an *opinion*, and I can *hear them all.* It's as if every sideline conversation at a basketball game is suddenly audible. That crashing tide of voices is heavy in a way that a stadium roar never can be. I realize, in a way that I never have before, what "a lot of people" means

It is a nightmare.

I feel crushed by the weight of this weird world.

I have no idea why I bother with this whole Net existence.

EDITOR'S NOTE: J. C. Herz spent a long time on-line. This is the final chapter of her book *Surfing the Net,* an account of her experiences in cyberland.

There's just too fucking much of this stuff, all the time, and it never stops.
All this information, it's toxic.
And Usenet's like some kind of continuous avalanche, tumbling in a loop.
I can't even think about IRC without getting queasy.
Jesus, I just want to shut off the crush of all these voices, the endless chatter, and
all the people that float right through me.
I'm sick of being a ghost. I feel myself starting to wear thin.
I'm sick of the overload,
sick of absorbing all this noise,
sick of cold coffee,
sick of the sleep deprivation,
sick of feeling strung out all the time,
sick of waking up in the morning with my brain ringing.
And you know, I really don't care if I never pick up another piece of goddamned
E-mail as long as I live.
I'm cooked.
I just can't *do* this anymore.
I am so
fucking
tired.

And I'm thinking, if I never log on again,
if this whole cyberplanet just vanished . . .
so what?

Net.death is starting to look pretty
liberating from where I sit
(to die, to sleep, perchance to dream . . .
Mmm. REM cycles).

Anyway, I could always come back to
Net.life, like Jason in *Friday the Thirteenth*.
God, it really is late, and my box of Cookie
Crisp is down to dust.

I think I'll draft a suicide note announcing
my impending Net Death >8–)

Maybe I'll become famous by helping
netaholics Net-kill themselves.

The "Kevorkian Virus" has a nice ring to it.
Muhahahaha.

No more hours of tapping into the web at 3
A.M.

No more flames.

Computers are a passing
"virus"; when they die out,
those among us who have
retained the ability to write,
think, and speak our minds
will again take over the world!

Marguerite J. Fraser
Courtney,
British Columbia

*Why, finally, if the new
technologies can run contrary
to the needs of human nature, do
we passively sit by as an entire
generation of children grows up
in cyberspace? At least part of
the explanation may be due to
the hype surrounding the digital
revolution almost from its
inception, a hype based largely
on the time-tested notion of
inevitability.*

MARK SLOUKA

195

No more horndog E-mail.

A decent night's sleep.

That's it, I'm slitting my virtual wrists :-)

Good-bye, people. It's been a kick.

But, as my man Dennis used to say,

I am OUTTA here.

NO CARRIER

———— ⟨▭◗ ————

We have this incredibly powerful machine called the computer but no problems to match it. Consider the dilemmas of 1995, and the computer's response. Is it curing cancer or AIDS? It may crunch numbers for researchers, but ultimately those cures will come from a scientist who's able to think about those diseases in an entirely new way, a breakthrough that requires the creativity of metaphor, something unique to the human mind. Is the computer fostering racial equality? No. Is it feeding starving Third World nations? No. Is it bringing peace between ethnic rivals? No. Keeping us safe from crime? No. Is it bringing American workers meaningful, well-paid jobs? If anything it's eliminating those jobs.

BRUCE BARCOTT
SEATTLE WEEKLY

VIRTUAL STUDENTS, DIGITAL CLASSROOM

NEIL POSTMAN

IF ONE HAS a trusting relationship with one's students (let us say, graduate students), it is not altogether gauche to ask them if they believe in God (with a capital G). I have done this three or four times and most students say they do. Their answer is preliminary to the next question: If someone you love were desperately ill, and you had to choose between praying to God for his or her recovery or administering an antibiotic (as prescribed by a competent physician), which would you choose?

Most say the question is silly since the alternatives are not mutually exclusive. Of course. But suppose they were—which would you choose? God helps those who help themselves, some say in choosing the antibiotic, therefore getting the best of two possible belief systems. But if pushed to the wall (e.g., God does not always help those who help themselves; God helps those who pray and who believe), most choose the antibiotic, after noting that the question is asinine and proves nothing. Of course, the question was not asked, in the first place, to prove anything but to begin a discussion of the nature of belief. And I do not fail to inform the students, by the way, that there has recently emerged evidence of a "scientific" nature that when sick people are prayed for they do better than those who aren't.

As the discussion proceeds, important distinctions are made among the different meanings of "belief," but at some point it becomes far from asinine to speak of the god of Technology—in the sense that people believe technology works, that they rely on it, that it makes promises, that they are bereft when denied access to it, that they are delighted when they are in its presence, that for most people it works in mysterious ways, that they condemn people who speak against it, that they stand in awe of it and that, in the "born again" mode, they will alter their lifestyles, their schedules, their habits and their relationships to accommodate it. If this be not a form of religious belief, what is?

In all strands of American cultural life, you can find so many examples of technological adoration that it is possible to write a book about it. And I would if it had not already been done so well. But nowhere do you find more enthusiasm for the god of Technology than among educators. In fact, there are those, like Lewis Perelman, who argue (for example, in his book, *School's Out*) that modern information technologies have rendered schools entirely irrelevant since there is now much more information available outside the classroom than inside it. This is by no means considered an outlandish idea. Dr. Diane Ravitch, former Assistant Secretary of Education, envisions, with considerable relish, the challenge that technology presents to the tradition that "children (and adults) should be educated in a specific place, for a certain number of hours, and a certain number of days during the week and year." In other words, that children should be educated in school. Imagining the possibilities of an information superhighway offering perhaps a thousand channels, Dr. Ravitch assures us that:

> in this new world of pedagogical plenty, children and adults will be able to dial up a program on their home television to learn whatever they want to know, at their own convenience. If Little Eva cannot sleep, she can learn algebra instead. At her home-learning station she will tune in to a series of interesting problems that are presented in an interactive medium, much like video games. . . .
>
> Young John may decide that he wants to learn the history of modern Japan, which he can do by dialling up the greatest authorities and

The future is green and low tech. We'll watch aquariums, not TVs.

A. Pavletich
Los Angeles,
California

So with a hundred "modern improvements"; there is an illusion about them; there is not always a positive advance. The devil goes on exacting compound interest to the last for his early share and numerous succeeding investments in them. Our inventions are wont to be pretty toys, which distract our attention from serious things. They are but improved means to an unimproved end.

HENRY DAVID THOREAU IN *WALDEN*

199

teachers on the subject, who will not only use dazzling graphs and illustrations, but will narrate a historical video that excites his curiosity and imagination.

In this vision there is, it seems to me, a confident and typical sense of unreality. Little Eva can't sleep, so she decides to learn a little algebra? Where does Little Eva come from? Mars? If not, it is more likely she will tune in to a good movie. Young John decides that he wants to learn the history of modern Japan? How did young John come to this point? How is it that he never visited a library up to now? Or is it that he, too, couldn't sleep and decided that a little modern Japanese history was just what he needed?

What Ravitch is talking about here is not a new technology but a new species of child, one who, in any case, no one has seen up to now. Of course, new technologies do make new kinds of people, which leads to a second objection to Ravitch's conception of the future. There is a kind of forthright determinism about the imagined world described in it. The technology is here or will be; we must use it because it is there; we will become the kind of people the technology requires us to be, and whether we like it or not, we will remake our institutions to accommodate technology. All of this must happen because it is good for us, but in any case, we have no choice. This point of view is present in very nearly every statement about the future relationship of learning to technology. And, as in Ravitch's scenario, there is always a cheery, gee-whiz tone to the prophecies. Here is one produced by the National Academy of Sciences, written by Hugh McIntosh.

School for children of the Information Age will be vastly different than it was for Mom and Dad.

Interested in biology? Design your own life forms with computer simulation.

Having trouble with a science project? Teleconference about it with a research scientist.

Bored with the real world? Go into a virtual physics lab and rewrite the laws of gravity.

These are the kinds of hands-on learning experiences schools could be providing right now. The technologies that make them possible are already here, and today's youngsters, regardless of economic status, know how to use them. They spend hours with them every week — not in the classroom, but in their own homes and in video game centers at every shopping mall.

I am typing this on a Remington Paragon. My uncle took this typewriter with him when he went away to college in 1926. I like the way it sounds and feels. It's a lot less impersonal than some computer would be; and it is a way to remember my uncle who died a few years ago.

Custis Whittock
Wallingford,
Pennsylvania

*This living flowing land
is all there is forever.
We are it
it sings through us
we could live on this Earth
without clothes or tools!*

GARY SNYDER

For many years my husband and I have been the summer guests at Walls' Apartments in Rehoboth Beach, Delaware. This sprawling, busy, and very prosperous summer

It is always interesting to attend to the examples of learning, and the motivations that ignite them, in the songs of love that technophiles perform for us. It is, for example, not easy to imagine research scientists all over the world teleconferencing with thousands of students who are having difficulty with their science projects. I can't help thinking that most research scientists would put a stop to this rather quickly. But I find it especially revealing that in the scenario above we have an example of a technological solution to a psychological problem that would seem to be exceedingly serious. We are presented with a student who is "bored with the real world." What does it mean to say someone is bored with the real world, especially one so young? Can a journey into virtual reality cure such a problem? And if it can, will our troubled youngster want to return to the real world? Confronted with a student who is bored with the real world, I don't think we can solve the problem so easily by making available a virtual reality physics lab.

The role that new technology should play in schools or anywhere else is something that needs to be discussed without the hyperactive fantasies of cheerleaders. In particular, the computer and its associated technologies are awesome additions to a culture, and are quite capable of altering the psychic, not to mention the sleeping, habits of our young. But like all important technologies of the past, they are Faustian bargains, giving and taking away, sometimes in equal measure, sometimes more in one way than the other. It is strange—indeed, shocking—that with the twenty-first century so close, we can still talk of new technologies as if they were unmixed blessings—gifts, as it were, from the gods. Don't we all know what the combustion engine has done for us and against us? What television is doing for us and against us? At the very least, what we need to discuss about Little Eva, Young John and McIntosh's trio is what they will lose, and what we will lose, if they enter a world in which computer technology is their chief source of motivation, authority, and, apparently, psychological sustenance. Will they become, as Joseph Weizenbaum warns, more impressed by calculation than human judgment? Will speed of response become, more than ever, a defining quality of intelligence? If, indeed, the idea of a school will be dramatically altered, what kinds of learning will be neglected, perhaps

rental establishment has been run very smoothly for over 40 years by Ms. Ann Walls McCool, a 77-years-young retired schoolteacher, who keeps track of guests, her properties, and rentals by using a lead pencil, index cards, and a "marble front" notebook. Although we have suggested that she might consider using more advanced technology, she responds that her pencil never goes "down."

Marguerite Donnelly
Ellicott City, Maryland 203

Loneliness may be the real disease of the next century, as we live alone, work alone, and play alone, insulated by our modem, our Walkman, or our television.

CHARLES HARDY

made impossible? Is virtual reality a new form of therapy? If it is, what are its dangers?

THESE ARE SERIOUS MATTERS, and they need to be discussed by those who know something about children from the planet Earth, and whose vision of children's needs, and the needs of society, go beyond thinking of school mainly as a place for the convenient distribution of information. Schools are not now and have never been largely about getting information to children. That has been on the schools' agenda, of course, but has always been way down on the list. For technological utopians, the computer vaults information-access to the top. This reshuffling of priorities comes at a most inopportune time. The goal of giving people greater access to more information faster, more conveniently and in more diverse forms was the main technological thrust of the nineteenth century. Some folks haven't noticed it but that problem was largely solved, so that for almost a hundred years there has been more information available to the young outside the school than inside. That fact did not make the schools obsolete, nor does it now make them obsolete. Yes, it is true that Little Eva, the insomniac from Mars, could turn on an algebra lesson, thanks to the computer, in the wee hours of the morning. She could also, if she wished, read a book or magazine, watch television, turn on the radio or listen to music. All of this she could have done before the computer. The computer does not solve any problem she has but does exacerbate one. For Little Eva's problem is not how to get access to a well-structured algebra lesson but what to do with all the information available to her during the day, as well as during sleepless nights. Perhaps this is why she couldn't sleep in the first place. Little Eva, like the rest of us, is overwhelmed by information. She lives in a culture that has 260,000 billboards, 17,000 newspapers, 12,000 periodicals, 27,000 video outlets for renting tapes, 400 million television sets and well over 500 million radios, not including those in automobiles. There are 40,000 new book titles published every year, and each day 41 million photographs are taken. And thanks to the computer, more than 60 billion pieces of advertising junk come into our mailboxes every year. Everything from telegraphy and photography in the nineteenth century to the silicon chip in the twentieth has

I miss library card catalogs terribly, and I hate searching for books using the on-line catalog. Reading books on a computer seems like having one's hand cut off (Ray Bradbury says that he wrote *Fahrenheit 451* in a library, and that he frequently dashed into the stacks to open books and smell them for inspiration.) My favorite bookstore owner has a habit of caressing his books as he hands them to the purchaser . . .

Perhaps analogous to the information highway are silk and plastic plants. They are pretty and they don't shed. They don't need watering, but neither do they grow and change, produce oxygen, or nourish animals.

Jill Carpenter
Sewanee, Tennessee

205

amplified the din of information intruding on Little Eva's consciousness. From millions of sources all over the globe, through every possible channel and medium—light waves, air waves, ticker tape, computer banks, telephone wires, television cables, satellites and printing presses—information pours in. Behind it in every imaginable form of storage—on paper, on video, on audiotape, on disks, film and silicon chips—is an even greater volume of information waiting to be retrieved. In the face of this we might ask, What can schools do for Little Eva besides making still more information available? If there is nothing, then new technologies will indeed make schools obsolete. But in fact, there is plenty.

One thing that comes to mind is that schools can provide her with a serious form of technology-education. Something quite different from instruction in using computers to process information, which, it strikes me, is a trivial thing to do, for two reasons. In the first place, approximately 35 million people have already learned how to use computers without the benefit of school instruction. If the schools do nothing, most of the population will know how to use computers in the next ten years, just as most of the population learns how to drive a car without school instruction. In the second place, what we needed to know about cars—as we need to know about computers, television and other important technologies—is not how to use them but how they use us. In the case of cars, what we needed to think about in the early twentieth century was not how to drive them but what they would do to our air, our landscape, our social relations, our family life and our cities. Suppose in 1946 we had started to address similar questions about television: What will be its effects on our political institutions, our psychic habits, our children, our religious conceptions, our economy? Would we be better positioned today to control TV's massive assault on American culture? I am talking here about making technology itself an object of inquiry so that Little Eva and Young John are more interested in asking questions about the computer than getting answers from it.

I am not arguing against using computers in school. I am arguing against our sleepwalking attitudes toward it, against allowing it to distract us from important things, against making a god of it. This is what Theodore Roszak warned against in *The Cult of Information:* "Like all cults," he wrote, "this one also has the intention of

After two computers and about seven years of my imagination reducing to the size of a video terminal, I am again writing with a fountain pen as I gaze at wider vistas . . . It is slightly slower in the desert but Voice Mail is here, fax is here, and TV from Death Star LA . . . And have we talked about the damn car? The absolute ruination of the West.

Diana Quinn Rose
Palm Springs,
California

PONA: PERSON OF NO ACCOUNT. *Anyone who does not have an account in cyberspace, anyone not on-line.*

I characterize myself as technologically out of step. Also, I wear a size 11 shoe, which, until recently, was not considered *au fait.*

Audrey Hargreaves
Los Angeles, California

enlisting mindless allegiance and acquiescence. People who have no clear idea of what they mean by information or why they should want so much of it are nonetheless prepared to believe that we live in an Information Age, which makes every computer around us what the relics of the True Cross were in the Age of Faith: emblems of salvation." To this, I would add the sage observation of Alan Kay of Apple Computer. Kay is widely associated with the invention of the personal computer, and certainly has an interest in schools using them. Nonetheless, he has repeatedly said that any problems the schools cannot solve without computers, they cannot solve with them. What are some of those problems? There is, for example, the traditional task of teaching children how to behave in groups. One might even say that schools have never been essentially about individualized learning. It is true, of course, that groups do not learn, individuals do. But the idea of a school is that individuals must learn in a setting in which individual needs are subordinated to group interests. Unlike other media of mass communication, which celebrate individual response and are experienced in private, the classroom is intended to tame the ego, to connect the individual with others, to demonstrate the value and necessity of group cohesion. At present, most scenarios describing the uses of computers have children solving problems alone; Little Eva, Young John and the others are doing just that. The presence of other children may, indeed, be an annoyance.

LIKE THE PRINTING PRESS before it, the computer has a powerful bias toward amplifying personal autonomy and individual problem-solving. That is why educators must guard against computer technology's undermining some of the important reasons for having the young assemble (to quote Ravitch) "in a specific place, for a certain number of hours, and a certain number of days during the week and year."

Although Ravitch is not exactly against what she calls "state schools," she imagines them as something of a relic of a pre-technological age. She believes that the new technologies will offer all children equal access to information. Conjuring up a hypothetical Little Mary who is presumably from a poorer home than Little Eva, Ravitch imagines that Mary will have the same opportunities as Eva "to learn any

For many people the process itself of computer knowledge and manipulation is what drives them. The ability to store, access, send and receive information for thousands of people is the end in itself, and many of these people are in the old-fashioned sense of illiteracy, very illiterate.

Robert Hamblin, M.D.
Santa Rosa, California

Traditional publishers may be elite, but software publishers are arrogant sheep.

MICHAEL MELLIN,
FORMER PUBLISHER
RANDOM HOUSE
ELECTRONIC
PUBLISHING

My mother is a NYC math teacher who chooses to use a pencil and paper to do all her work. You can imagine how much faster it would be if she used a calculator to add thirty numbers and then ✐➯

subject, and to learn it from the same master teachers as children in the richest neighbourhood." For all of its liberalizing spirit, this scenario makes some important omissions. One is that though new technologies may be a solution to the learning of "subjects," they work against the learning of what are called "social values," including an understanding of democratic processes. If one reads the first chapter of Robert Fulghum's *All I Really Need to Know I Learned in Kindergarten,* one will find an elegant summary of a few things Ravitch's scenario has left out. They include learning the following lessons: Share everything, play fair, don't hit people, put things back where you found them, clean up your own mess, wash your hands before you eat and, of course, flush. The only thing wrong with Fulghum's book is that no one has learned all these things at kindergarten's end. We have ample evidence that it takes many years of teaching these values in school before they have been accepted and internalized. That is why it won't do for children to learn in "settings of their own choosing." That is also why schools require children to be in a certain place at a certain time and to follow certain rules, like raising their hands when they wish to speak, not talking when others are talking, not chewing gum, not leaving until the bell rings, exhibiting patience toward slower learners, etc. This process is called making civilized people. The god of Technology does not appear interested in this function of schools. At least, it does not come up much when technology's virtues are enumerated.

The god of Technology may also have a trick or two up its sleeve about something else. It is often asserted that new technologies will equalize learning opportunities for the rich and poor. It is devoutly to be wished for, but I doubt it will happen. In the first place, it is generally understood by those who have studied the history of technology that technological change always produces winners and losers. There are many reasons for this, among them economic differences. Even in the case of the automobile, which is a commodity most people can buy (although not all), there are wide differences between the rich and poor in the quality of what is available to them. It would be quite astonishing if computer technology equalized all learning opportunities, irrespective of economic differences. One may be delighted that Little Eva's parents

divide them to obtain an average. She wants to keep her mind sharp—and sharp it is!

Laurie Gold
Woodbury, New York

I just don't personally believe in reading novels on a computer screen. There's a lot of content that's now being delivered on paper that's fine on paper.

OLAF OLAFSSON,
PRESIDENT
SONY ELECTRONIC
PUBLISHING

Why do they call them personal computers anyway? Does my computer ever listen to my problems? Will it attend my funeral?

Ann McMartin
Camarillo, California

could afford the technology and software to make it possible for her to learn algebra at midnight. But Little Mary's parents may not be able to, may not even know such things are available. And if we say that the school could make the technology available to Little Mary (at least during the day), there may be something else Little Mary is lacking.

It turns out, for example, that Little Mary may be having sleepless nights as frequently as Little Eva but not because she wants to get a leg up on her algebra. Maybe because she doesn't know who her father is, or, if she does, where he is. Maybe we can understand why McIntosh's kid is bored with the real world. Or is the child confused about it? Or terrified? Are there educators who seriously believe that these problems can be addressed by new technologies?

I do not say, of course, that schools can solve the problems of poverty, alienation and family disintegration, but schools can *respond* to them. And they can do this because there are people in them, because these people are concerned with more than algebra lessons or modern Japanese history, and because these people can identify not only one's level of competence in math but one's level of rage and confusion and depression. I am talking here about children as they really come to us, not children who are invented to show us how computers may enrich their lives. Of course, I suppose it is possible that there are children who, waking at night, want to study algebra or who are so interested in their world that they yearn to know about Japan. If there be such children, and one hopes there are, they do not require expensive computers to satisfy their hunger for learning. They are on their way, with or without computers. Unless, of course, they do not care about others or have no friends, or little respect for democracy or are filled with suspicion about those who are not like them. When we have machines that know how to do something about these problems, that is the time to rid ourselves of the expensive burden of schools or to reduce the function of teachers to "coaches" in the uses of machines (as Ravitch envisions). Until then, we must be more modest about this god of Technology and certainly not pin our hopes on it.

We must also, I suppose, be empathetic toward those who search with good intentions for technological panaceas. I am a teacher myself and know how hard it is to contribute to the making of a

People go into book publishing and bookselling because they want to be involved with words and books. They don't want to make money. If they do, they've made a serious mistake. The CD-ROM people, and the gaming community—those people are in love with games, in love with computers, and in love with making money. They have no problem talking about units.

MICHAEL LYNTON,
FORMER SENIOR VICE
PRESIDENT
WALT DISNEY COMPANY
PUBLISHING DIVISION

213

I hope this is the 10,000th letter of appreciation that reached you because of your manifesto in Sunday's *L.A. Times*. . . . This typewriter—used daily—is 75-years-old. Uses not a drop of electricity.

Joe Scalzo
Sierra Madre,
California

civilized person. Can we blame those who want to find an easy way, through the agency of technology? Perhaps not. After all, it is an old quest. As early as 1918, H. L. Mencken (although completely devoid of empathy), wrote, "There is no sure-cure so idiotic that some superintendent of schools will not swallow it. The aim seems to be to reduce the whole teaching process to a sort of automatic reaction, to discover some master formula that will not only take the place of competence and resourcefulness in the teacher but that will also create an artificial receptivity in the child."

Mencken was not necessarily speaking of technological panaceas but he may well have been. In the early 1920s a teacher wrote the following poem:

Mr. Edison says
That the radio will supplant the teacher.
Already one may learn languages by means of Victrola records.
The moving picture will visualize
What the radio fails to get across.
Teachers will be relegated to the backwoods,
With fire-horses,
And long-haired women;
Or, perhaps shown in museums.
Education will become a matter
Of pressing the button.
Perhaps I can get a position at the switchboard.

I do not go as far back as the radio and Victrola, but I am old enough to remember when 16-millimeter film was to be the sure-cure. Then closed-circuit television. Then 8-millimeter film. Then teacher-proof textbooks. Now computers.

I know a false god when I see one.

On the eve of the 21st century, what basis do we have for knowing if anything on our computer screens is really error-free? The complexity of today's machines is far beyond our ability to check them by hand. All the fine points are floating; the only orthodox solution is to have more faith. We must accept on faith that our telephone bills are absolutely error-free just as we must accept the integrity inside a microprocessor.

Absolute faith is a software issue, a sliver of silicon surrounded by a sea of the faithful, their arms outstretched, waiting for instructions from a ROM (read only memory). It can be read but not changed, just like the Bible, humanity's original ROM.

JOHN HOCKENBERRY

Rapture and Redemption in the Virtual World

MARK SLOUKA

AMERICAN CULTURE periodically suffers bouts of religious fervor the way a malaria victim suffers ague. This is not new. It is our fate, and, considered all in all, we bear it well. What *is* new, however, is the form the symptoms take. As each new age (extending the metaphor a bit), develops its own brand of quinine, the spirit adjusts. In 1735 we trembled over Jonathan Edwards' fiery pit. In 1801 we frothed at the mouth and spoke in tongues around the campfires at Cane Ridge, Kentucky. In 1858, attending prayer lunches at noon during the Third Great Awakening, we turned Wall Street into a ghost town. In a town in Texas, more recently, we blocked traffic with our Winnebagos, gazing at a roadside billboard rumored to be hiding the features of the Savior in a forkful of spaghetti. Even those aware of history, it seems, are doomed to repeat it.

With a difference, though. This time around, the spirit has returned in a new form: think of it as a kind of genetic mutation, an unexpected anomaly—tougher, more enduring than its predecessors. This time the spirit is playing for keeps. Its goal is still nothing more or less than the age-old dream of transcendence. This time, however, technologically sophisticated, funded to the teeth, it may well win. Already the Heavenly City shines in the eyes of the digitally devout. Immaterial, perfect, it will soon be open for occupancy. Just in time for the millennium.

Overstatement? Hyperbole? According to John Perry Barlow, Grateful Dead lyricist-turned-computer-cowboy, what the computer world is doing, "is taking material and making it immaterial: Now is the flesh made word, in many respects." According to Michael Benedikt of the University of Texas, cyberspace will allow us to make up for the Fall from Eden, to redeem ourselves in God's graces; having shed "the ballast of materiality," we will gain the Heavenly City at last. "Reality is death," he intones, soberly. Bruce Mazlish of MIT, fellow of the American Academy of Sciences, takes it a bit further. Human beings, he notes, disgusted by the body, tired of being ignored by the world of objects, repulsed by the irrational, hideous world, are turning to virtual technologies because "they do not wish to settle for such a limited condition. They aspire to be angels, if not God." Catching the spirit, Nicole Stenger of the University of Washington assures us that soon we will all be "rocking and humming in televirtuality," members of a community of "highly unstable, hermaphrodite angels." For his part, Kevin Kelly, associate editor of *Wired* magazine, seems ready to thrash in the aisles. "I can't imagine anything more intoxicating," he writes, "than playing God."

Faced with all this, one gropes for an appropriate response. Embarrassment? Disbelief? Surely, we think, when MIT professors and IBM

217

researchers rhapsodize about the possibilities of engineering a synthetic environment, or technologizing our "surround" to the point of hallucination, or ascending to the Heavenly City via fiber-optic cable, or "uploading" the human consciousness into RAM ("transcending the meat," as they so neatly put it) they're speaking metaphorically. Right? Alas, no. "The only thing wrong with the universe," says former Apple employee Ken Karakotsios, summing up the general mood nicely, "is that it is currently running someone else's program."

Mr. Karakotsios, of course, is onto something. What makes the self-described "net-religionists" worth watching is that they've managed to marry the yearning for transcendence to a potential 3.5 *trillion* dollar industry. In short, they're serious. Unabashedly messianic, surprisingly market-wise, formidably well-funded, they're grafting the tradition of millennial dreaming onto both the not-so-free market and a powerful new technology. In short order, they'll be in a position not only to sell the age-old dream of the Heavenly City, but to rent out space there.

Absurd? Already, the *New York Times* informs us, the first communication links between the human nervous system and the silicon chip have been established. Within a generation, perhaps less, virtual systems will enable us to "shed the ballast of materiality" altogether, to enter, through a kind of digital rapture, what the enthusiasts call "a fully englobing environment." Soon, in other words, the primitive illusions already available on CD-ROM will be indistinguishable from the real thing. Instead of merely planting and harvesting virtual tomatoes on Forever Growing Garden, you'll be able to taste them.

But let's get serious. Beyond the millennial rhetoric, beyond the high-tech jargon, beyond that Orwellian and seemingly ubiquitous adjective, virtual, is a marketing scheme of unrivaled audacity, unprecedented scope, and nearly unimaginable impact: a scheme, that is, designed to sell us copies of the thing we already have available to us for free—life itself. "I believe," says the computer-assisted design specialist John Walker, that "in the fullness of time, every object in the world, manufactured or not, will be modeled inside a computer. This is a very, very big market. This is everything."

The fullness of time may be nearer than we think. Soon, writes Bill Gates in

I am a glass blower and the basic tools I use every day were used 2,000 years ago and have changed very little. So to me a pencil and pencil sharpener are high tech.

Michael Kraatz
Canaan, New York

To paraphrase Brendan Behan, an Irishman without a computer, I see nothing about this superhighway that makes the roads safer or keeps old people warmer in the winter, so there's no use for it.

P. Sean Clark
El Segundo, California

It is true that in a general sense the Luddites were not successful either in the short-run aim of halting the detestable machinery or in the long-run task of stopping the Industrial ✏️➤

The Road Ahead, "you will be able to conduct business, study, explore the world and its cultures, call up any great entertainment, make friends, attend neighborhood markets, and show pictures to distant relatives—without leaving your desk or armchair . . . your network connection . . . will be your passport into a new, mediated way of life."

The mediated life, of course, aided by one of the great migrations of human history—the movement inside our own homes—is already here. As more of the hours of our days are spent in synthetic environments, partaking of electronic pleasures, life itself is turned into a commodity. As the natural world fades from our lives, the unnatural one takes over; as the actual, physical community wanes, the virtual one waxes full and fat. Bill's plan (and he's not alone), is to take advantage of the social momentum. He wants a piece of the action. The new world, he promises, will be "a world of low-friction, low-overhead capitalism, in which market information will be plentiful and transaction costs low." What he neglects to mention is that (just as in the Annie Liebovitz photograph which graces the jacket cover of his book), the road to "shopper's heaven" leads past him, and that he happens to be manning the tollbooth.

Bill's vision of a "friction-free" world, one must admit, has a certain Singaporean charm. From forcing people to "turn down their volume," to proposing that we apply something like the Motion Picture Association's movie ratings to social discourse, to suggesting that a virtual forest of hidden surveillance cameras be installed "to record most of what goes on in public," Bill is out to make the world free from friction and safe for commerce. What he seems to have overlooked (there's no way to put this delicately) is that friction, in social life as in the bedroom, has its virtues. The "friction" he would spare us, after all (and here his voice merges with those of the silicon seraphim quoted earlier), is the friction of personal contact, of physical movement, of social interaction. Cultural life, one wants to remind him, requires friction. As does democracy.

It's always possible, of course, that democracy, or a thriving social life, are not what Bill and his fellow enthusiasts are after because they truly believe that these notions (like sex, or physical space), will be

Revolution and its multiple miseries; but that hardly matters in the retrospect of history, for what they are remembered for is that they resisted, not that they won. Some may call it foolish resistance ("blind" and "senseless" are the usual adjectives), but it was dramatic, forceful, honorable, and authentic enough to have put the Luddites' issues forever on record and made the Luddites' name as indelibly a part of the language as the Puritans'.

KIRKPATRICK SALE

221

My husband, Sol, and I are artists of the non-technical variety, who view the encroachment of computers into what's left of human expression as a seductive form of brain death.

Sharon Rothman
Bogota, New Jersey

the vestigial limbs of the brave new age, exhibited by a handful of Luddites and die-hard humanists, and no one else. (In the digital future, Nicole Stenger of the University of Washington reminds us, "cyberspace will be your condom.") It's possible, as well, that these latter-day Nathan Hales really believe in the "liberty" of electronic shopping, of being able, as Gates promises, to instantly order the "cool" sunglasses Tom Cruise wears in "Top Gun" while watching the movie. It's possible, finally, that it is simple naivete that has the technoevangelists whistling past the authoritarian graveyard (as they usher in Orwell's Panopticon), and not some fellow feeling for those buried within.

Whether they really believe in their brave new world, however, is ultimately beside the point. They're building it. And in the friction-free future, jacked into paradise, we'll have the "liberty" of living (or rather, of buying the *illusion* of living), through the benevolent offices of a middleman as nearly omnipotent as god himself. Freedom? A more perfect captivity is difficult to imagine.

AFTERWORD

DORIS GRUMBACH

I write a final footnote to all the essays, letters and articles printed here, and after the news of the Leaddite movement has spread to a receptive world. Since the time that Bill Henderson extracted from a book of mine the idea that there could be a serious pothole on the communications highway and called me a founder of the Lead Pencil Club (only friendship could have made that jump), I have wrestled with the growing intrusion, into my writing life, of the monster I had hitherto excoriated.

I noticed that when Otto Friedrich, prolific journalist and editor for *Time*, died in 1995, his obituary boasted of the fact that he had gone on writing his more-than-forty cover stories and hundreds of shorter pieces on "an old-fashioned Royal typewriter." *Time* "gave him a special dispensation to continue using [it] long after it converted to computers." To my shame, I had none of this fidelity to an old and trusty Underwood, having followed the madding crowd to an electric typewriter and then (*mea culpa*) took up with a word processor and its companion printer. I contend it is to my credit that I do not *write* on the . . . the *thing,* however, but use it merely as editor, typist, and duplicator. The salesman who sold it to me rightly termed it a clone, and spoke of what I hitherto thought applied to wrenches and screwdrivers: *hardware.*

I am opposed to other technological advances: the many-voiced and multi-function telephone and the answering machine. I hate and resent them both, the machine that can make a caller wait and listen to music; the disembodied voices on a tape and cutesy messages; the calls you can make to it from another location.

There is a little man who resides in my new answering machine who tells me how many messages there are and provides me with other information of no use to me. Most of the calls he reports I never wished to know about anyhow. I am nostalgic for the time when I was ignorant of the calls I received and so was not required to re-call persons I did not wish to speak to in the first place. Right now, I am trying to figure out how to get him out of that black box so he can cook dinner and empty the trash.

The operative word in our society now is "communications." Universities have whole departments for that dubious activity and grant degrees in it. There are chains of stores devoted to selling machines and "programs" (rightly termed "software") to facilitate communicating with everyone, even internationally, in the flash of a stroke on the keyboard.

There are a million ways of communicating at a time when there is far less of interest and importance to communicate. I learned a valuable lesson from contemporary television. With fifty channels available to me, I find less to look at than when there were four. I want and need very little of it.

There is no one I wish to communicate with that cannot be reached by a handwritten letter at the cost of thirty-two cents, one way. It is a reliable, slow method of telling someone something, and I can wait for an answer in the same mode.

We are still reading a book my fellow-founder (emeritus), Henry David Thoreau, wrote more than one hundred and fifty years ago with, I believe, the lead pencil his father had manufactured. We will, I ardently hope, *still* be reading *Walden* when all the communications held, forwarded, and recorded, all the faxes and Internet and Online communications on screens, spewed out and responded to, *instanter,* have dissipated into oblivion, or will soon.

Thoreau observed that we do not ride on railroads; they ride on us. Machines do not facilitate, they make slaves of us. We

As a professional archivist, I look with great discomfort on the current scramble toward the artificial light of information technology . . . Americans seem to have a great desire to isolate themselves from each other, and the so-called information highway will help them to do this better than before. Just as the freeways built after WWII helped destroy what was once a thriving urban culture, so our new communication toys will help to explode centers of culture like universities. And the more isolated people become, the easier it will be to manipulate them.

Christian Class
San Francisco,
California

225

buy them because they are out there, and we use them because they promise to make a difference in our personal lives.

But of course they make very little significant difference. They occupy more time than they are worth. When electricity fails, as it does often in Maine where I live, we are powerless, because, without the use of the telephone, the answering machine, the computer and the printer and, oh yes, the electric toothbrush, we are persons sitting in the dark, clones of our former selves.

—— ✏ ——

I think sadly about the 30,000 Alpine carrier pigeons who will be "discharged" from the Swiss army next year.

They cost too much, see. About $465,000 a year. Besides, the Swiss army figures it's time to up-date the old communications system.

Carrier pigeons have been ferrying important messages for the Swiss army for 77 years. Flying as high as 6,000 feet and as fast as 50 mph, the birds also carry microfilm and tape recordings. . . .

On the other hand, you can't punch a pigeon and get the latest sports scores, weather forecast, or Janet Jackson's psychic.

You can't use a pigeon to report a stolen credit card or move $10,000 from one mutual fund to another or order a Rolex watch from an exclusive catalog.

No, I don't suppose there is a single vehicle on the whole information superhighway that is compatible with a carrier pigeon.

STEPHANIE SALTER

BOOK REPORT

When we founded the Lead Pencil Club, it was hard to find any support for our position in print. This has changed, thank goodness. Here are a few of the best titles, and two to avoid. All our recommended titles are jargon and hot-air free, and they are written by authors who indeed seem to have a life.

Neil Postman's TECHNOPOLY: THE SURRENDER OF CULTURE TO TECHNOLOGY (Vintage paperback, 1993) sums up clearly the forces that have arrived in this century and are promising to take us over in the next: "Stated in the most dramatic terms, the accusation can be made that the uncontrolled growth of technology destroys the vital sources of our humanity. It creates a culture without moral foundation."

Another trustworthy guide is Clifford Stoll, a computer jock with twenty years at the keyboard. His SILICON SNAKE OIL: SECOND THOUGHTS ON THE INFORMATION HIGHWAY (Doubleday, 1995) should be consulted without fail by all concerned readers, from PONAs to Cybergeeks and all caught in the baffled middle. Stoll has been to nerdsville and returned into the sunlight to discover real life with his young daughter.

Stoll warns that untold "Zillions of dollars" have been spent for computer equipment that most of us don't need. He compares the Post Office's snail mail with E-mail and discovers that the Post Office is far more reliable and cheaper. He reveals that Internet postings are packed with utter drivel, and finds that Internet flame wars are "reminiscent of Ostrogoths and Visigoths."

Home shopping by modem? "Look at the detritus, dross, and dreck sold on TV shopping channels and note that the same people will own the Internet shopping networks."

Paperless office? Forget it. Stoll says computers generate more paper than the systems they replaced. The computer as a tool for thinking? "Ouch, we need a tool to spare us the effort of thinking?"

Stoll is devastating about computers and education: "Kids learn to stare at a monitor for hours on end and accept what a machine says without arguing . . . the world is a passive, preprogrammed place, where one click on the mouse gets the right answer."

Read the book. Stoll is a technophile you can trust. A paperback edition should be out any day now from Anchor.

For readers interested in a true account of the lurking horrors behind all the electronic bulltwaddle, Mark Slouka's WAR OF THE WORLDS: CYBERSPACE AND THE HIGH-TECH ASSAULT ON REALITY (Basic, 1995) will serve as a convincing reason to drive a spike through your computer's heart.

Slouka's book is about a small but influential group of self-proclaimed

"Net-religionists" and "wannabe gods" who believe the physical world can and should be delivered into a computer and that the future of mankind is not in RL (real life) but in VR (virtual reality). According to Slouka these folks are working very hard and spending enormous amounts of federal and private money to engineer their version of the apocalypse. "When a significant number of powerful individuals—scientists, academics, authors, engineers, computer programmers—following the scent of a potential $3.5 trillion industry begin referring to the human body as *meat* (the expression is a common one among the digerati), its time for those still foolishly attached to theirs to start paying attention . . ."

Unlike this writer, who is fast losing his sense of humor in the face of this cyberbully onslaught, Slouka keeps his:

Every other week someone says that books are dead or dying, that just around the corner is the black hour when they will be curiosities like stereopticon slides or milk stools—probably the same thing they said when radio was invented, when television flickered its way into our living rooms.

E. ANNIE PROULX

Yellow pencils are the taxicabs of our thoughts.

Bob Marvin
Eustis, Maine

Now we chatter anonymous and alone in a darkened room, behind a scrim of cursors and bytes, watching the electronic fire burn. We inhabit a place with 500 channels and nobody home. It hardly seems like progress.

BOSTON GLOBE

"... the cyberists' self-described in-your-face vision can be quite entertaining ... At times, in fact, it offers as much fun—there's no way to put this delicately—as some of the aphorisms of Pol Pot in the 1970s or Hitler's more memorable pronouncements in MEIN KAMPF."

Call Basic Books and order a copy. Don't assume a fetal position and pray these people will vanish into their own VR. Follow Slouka's advice and head them off with common sense and a bit of decency.

Here are other books you should consult:

THE GUTENBERG ELEGIES by Sven Birkerts (Faber and Faber, 1995). Contemplates the future of books and reading. Wonderfully written.

SURFING THE INTERNET by J .C. Herz (Little, Brown, 1995). A young woman's adventures in Internet sex and chat, and her ultimate disgust with the whole affair. A detailed report from the front.

THE EMPEROR'S VIRTUAL CLOTHES by Dinty Moore (Algonguin, 1995). Another report from the front—balanced and thoughtful.

RESISTING THE VIRTUAL LIFE edited by James Brook and Iain A. Boal (City Lights, 1995). Essays by a variety of folks from Lawrence Ferlinghetti's stellar small press.

Books to avoid—Nicholas Negroponte's BEING DIGITAL (Knopf, 1995). Pure vaporware. I read the entire thing and remember nothing except the following: Negroponte does not like to read (he is dyslexic). He predicts we will be able to handle the garbage overflowing from the Infobahn with a "personal agent" (another computer, of course) that will be programmed with our personality and select what we might like from the info glut. (Your mind and soul will be in the can, so to speak). He also predicts lots of people will be out of work when the digital revolution really takes hold, and that's too bad. Other than that he's so vague that he's impossible to sum up. Oh yes, he dines with presidents, kings, and heads of Japanese electronic firms, and he runs MIT's Media Lab. (A vapor site, I assume.) I learned this from his book.

Bill Gates' best seller The Road Ahead (Penguin USA, 1995) isn't half as bad as Negroponte's. After reading it you can actually remember a few things he predicts: a wallet PC for instance, with 1,000 photos of your kid and Global

I have to confess when I was younger and worked in the electronic industry, I thought how great, how marvelous! And I believed the technology had great promise. But I have changed my mind with a vengeance! Instead of progressing in an orderly manner with some thought to down the road problems, it has raced ahead willy-nilly to nothing.

D. Holland
Sonoma, California

231

On the Internet, there are few editors to check and either affirm or deny our vision. There are few readers, other than friends, to count my influence—good or bad. It is a shame but it is also true, I think, that fiction, poetry, and essays on the Net are heading toward the status of junk mail.

W. SCOTT OLSEN
AWP CHRONICLE

Positioning System installed, or how nifty his zillion dollar new house is; or how new computer technology will eliminate distance and bring the marketplace and the movie theater right into your living room so you don't have to ever leave. He promises that computers won't replace teachers in the classroom—but will be pivotal to education. In fact, the scariest idea in this book isn't there at all—it's Gates' promise that he will contribute all his royalty proceeds to a foundation to study the uses of electronic gadgets in the classroom. Guess who will benefit when that foundation finds millions of nifty applications? You can bet it won't be the teachers or kids.

Gates is a gee-whiz lad with a boyish grin who is frozen at age 15. He's stumbled onto something, and he hasn't got a clue what he's done or what it's for. But he's utterly determined to guard the play fort he has fabricated. All of this *is* a game, you see?

BH

The god of our godless age is speed. Driven by our obsession to compete, we have embraced this electronic god with a frenzy.

Soon, blessed with fax, voice and E-mail, computer hookups and TVs with hundreds of channels, we won't have to leave our lonely rooms—not to write a check, work, visit, shop, exercise or make love (virtual reality will serve.)

We won't write letters to friends because we won't have friends—just electronic anachronisms. Next century, nobody will know the meaning of "love," though "self-love" may survive.

We will have raced at incredible speeds on the information superhighway to reach our final destination—Nothing.

Our Director Emeritus, Henry David Thoreau, said it first in 1844: "We are in great haste to construct a magnetic telegraph from Maine to Texas, but Maine and Texas, it may be, have nothing important to communicate."

Like Henry David Thoreau, our Club is cranky. We honor provocation. We want to start an argument in the broad flapping American ear.

With Founding Director Doris Grumbach, we like to imagine that computers had "contracted a fatal illness for which there is no cure."

Politicians predict that the information superhighway will bring people closer together, but the Club asks: Now that voice mail answers

233

your telephone calls, are we more in touch or are we drifting further apart?

Our schools are hustled with the hype that they must invest millions of dollars in computers that will eventually replace teachers in many classrooms. Caught in the machine madness, our educators too soon forget that only one human being can teach another.

The Lead Pencil Club warns that we have gone Gizmo Gaga. We ask where are we zapping to on this highway? And shouldn't some scouts be sent out ahead to survey the terrain before we all accelerate?

Our pledges: We will avoid fax and hang up on voice mail. We will receive no E-mail and send none. If our computers develop a virus, we will seek no cure. Our communications will be face to face. If direct human contact is not possible, we will write letters in our own handwriting because that handwriting is a mark of our personality.

Of course, in all our correspondence, we will favor the lead pencil — simple, erasable, light, portable, and responding immediately to the mind, that quirky little expendable that the superhighway would like to forget as it rushes past on its way to oblivion.

What's your hurry? Speed Kills. Where's the Fire? Haste Makes Waste. Back to Basics. Not So Fast!

As Henry David Thoreau said, "Simplify."

Leadites Unite!

235